SpringerBriefs in Molecular Science

History of Chemistry

Series Editor

Seth C. Rasmussen, Fargo, USA

For further volumes:
http://www.springer.com/series/10127

David E. Lewis

Early Russian Organic Chemists and Their Legacy

 Springer

David E. Lewis
Department of Chemistry
University of Wisconsin-Eau Claire
Eau Claire WI 54702-4004
USA

ISSN 2191-5407   e-ISSN 2191-5415
ISBN 978-3-642-28218-8   e-ISBN 978-3-642-28219-5
DOI 10.1007/978-3-642-28219-5
Springer Heidelberg New York Dordrecht London

Library of Congress Control Number: 2012931860

Printed on acid-free paper

Springer is part of Springer Science+Business Media (www.springer.com)

Preface

The organic chemists of Russia during the pre-revolutionary period counted among their number some of the most creative and talented chemists of the nineteenth and early twentieth centuries, as is attested by the number of reactions and empirical rules bearing their names. From quite modest beginnings, higher education in Russia gradually grew into the point where Russian universities and the organic chemists in them could produce work to rival any done in France, Germany, Britain, or America. In this Brief, the history of the development of organic chemistry in Russia is discussed, with special emphasis on the Russian organic chemists who made important contributions to the science.

Acknowledgments

It is a very real pleasure to acknowledge the support of colleagues in the Division of the History of Chemistry of the American Chemical Society for the past two decades. Without their continued encouragement and occasional criticism, and especially without the official organ of the Division, *The Bulletin for the History of Chemistry*, much of my research into the development of organic chemistry in Russia would never have been carried out. The Division and the *Bulletin* both provide an essential outlet for scholarly work in the history of chemistry.

I am also indebted to the University of Wisconsin–Eau Claire for the award of a sabbatical leave, during which most of this manuscript was prepared and completed.

My family has been a constant source of support and love for me during the writing of this book, and I am grateful to them for their forbearance. They are truly inspirational.

Contents

Spelling and Transliteration

Russian uses the Cyrillic alphabet, and so names must be transliterated to the Roman alphabet. The exact transliteration used depends on the language into which the transliteration occurs, and even this is not a constant within the same language. A good example of this is provided by the name of N. N. Sokolov, which is transliterated into German as Socoloff by *Justus Liebigs Annalen der Chemie*, and as Sokoloff by *Erdmans Journal für Praktische Chemie*.

In this book, the BGN/PCGN romanization system for Russian is used. The consonants ж, ц, ч, ш and щ are transliterated as zh, ts, ch, sh, and shch respectively. The vowels й, ы, э, ю, and я are transliterated as i, y, e, yu, and ya, respectively. The vowel e at the beginnings of words is transliterated as ye. The soft sign (ь) is rendered as ′, and the hard sign (ъ) is rendered as ″.

In citations of articles in western journals, names are given as transliterated by the journals, so the name of one individual often appears with more than one spelling.

Chapter 1
The Evolution of Higher Education in Russia

1.1 Introduction: The Rulers of Imperial Russia, 1682–1917

Understanding the effects of politics on the development of higher education in Russia is one key to understanding the development of organic chemistry in that nation. Since secular education in Russia went from high levels of autonomy to rigid central control as the monarch and policies changed, it behooves us to briefly look at the political events of the nineteenth century, and to see how these events affected the growth of the universities in Russia. As we shall see, this centralization of power in education tended to swing wildly from being very beneficial to the universities, permitting the innovations that would facilitate to the development of organic chemistry, to being a major impediment to innovation.

Table 1.1 lists the rulers of Imperial Russia from Peter the Great until Nicholas II [1]. It is worthwhile noting that of Peter the Great's immediate successors, only his wife, Catherine I, continued his policies of modernization. Both Peter II and Anna were reactionary monarchs who had little empathy with Peter's vision. The infant Tsar, Ivan IV, was deposed by Elizabeth before he had been on the throne much more than a year, and spent the remainder of his short life under imprisonment. The next two Empresses, Elizabeth and Catherine the Great both concurred in Peter's vision for a modernized Russia, and worked to bring that vision to reality. After Catherine, her son, Paul I (whom she tried to prevent ascending to the throne), became Tsar for a brief period of time before he was, in turn, succeeded by his son, Alexander I.

The nineteenth century was an eventful time for Russian higher education. The reign of Alexander I, the first of the Tsars of the new century, was a study in contrasts: he began as a liberalizer, and in the first three years of his reign he declared new universities in four cities of the empire. Under the University Statute of 1804, the new universities were established according to the German, rather than the French model. At the head of each university was a council of professors, which elected the rector, named professors to chairs, determined the course of

D. E. Lewis, *Early Russian Organic Chemists and Their Legacy*,
SpringerBriefs in History of Chemistry, DOI: 10.1007/978-3-642-28219-5_1,
© The Author(s) 2012

Table 1.1 Rulers of Imperial Russia, 1682–1917

Ruler	Reign
Peter I, the Great (Пётр I Алексеевич Романов, Пётр I, or Пётр Великий)	1682–1725
Catherine (Екатерина I Алексеевна)	1725–1727
Peter II (Пётр II Алексеевич)	1727–1730
Anna (Анна Иванова)	1730–1740
Ivan VI (Иван VI Антонович)	1740–1741
Elizabeth (Елизавета Петровна)	1741–1762
Peter III (Пётр III Фёдорович)	1762
Catherine II, the Great (Екатерина II Великая)	1762–1796
Paul I (Павел I Петрович)	1796–1801
Alexander I (Александр I Павлович)	1801–1825
Nicholas I (Николай I Павлович)	1825–1855
Alexander II (Александр II Николаевич)	1855–1881
Alexander III (Александр III Александрович)	1881–1894
Nicholas II (Николай II Александрович Романов)	1894–1917

studies, and acted as the academic council and university court of highest instance. Each university was the center of a school district; the universities directed the work of primary and secondary educational institutions and fulfilled the functions of censors. In the last half of his reign, Alexander became much more arbitrary and reactionary, seeking to repeal many of the liberal reforms of the early years of his reign.

His son, Nicholas I is often described as the most reactionary of Russian monarchs; his reign began with the suppression Decembrist revolt, and continued by exercising censorship and control over all aspects of Russian life—including the universities. In 1833, the minister of education, Uvarov,[1] devised a program known as "Orthodoxy, Autocracy and Nationality," whose principles were stated by Uvarov in the following terms in a circular to educators under his ministry:

> It is our common obligation to ensure that the education of the people be conducted, according to Supreme intention of our August Monarch, **in the joint spirit of Orthodoxy, Autocracy and Nationality**. I am convinced that every professor and teacher, being permeated by one and the same feeling of devotion to the throne and fatherland, will use all his resources to become a worthy tool for the government and to earn its complete confidence [2].

Uvarov was a classicist of international renown, so much of his effort was focused on the classics and their teaching. He spent much less time concerned with the physical sciences, which may have been to their benefit. Uvarov did, however,

---

[1] Sergei Semënovich Uvarov (Сергей Семёнович Уваров, 1786–1855) was a classical scholar who rose to become an important advisor to Tsar Nicholas I. He was well connected with the royal family: his godmother was Catherine the Great. From 1811 to 1822, he was Curator of the St. Petersburg educational district, and in 1832 he was appointed Deputy Minister of Education. The following year, he succeeded his father-in-law, Count Razumovskii, as minister. He was elected to the Academy of Sciences in 1811, and was its President from 1818 until his death.

recognize that the teaching of the physical sciences lagged behind western Europe, and did recognize that Russia did not have the native professors capable of providing the needed advanced education. Thus, in a truly ironic twist of fate, the same individual who devised and promulgated the very reactionary "Orthodoxy, Autocracy and Nationality" policy was also responsible for reinstating the *komandirovka*, a study abroad program that permitted scientists in Russia to pursue advanced education in Western Europe. Viewed from the perspective of history, no other policy was as influential in building a Russian professoriate in chemistry over the course of the nineteenth century. Thus, despite being a truly reactionary monarch, Nicholas I actually oversaw an improvement in the quality (if not the quantity) of Russian higher education—especially in the physical sciences.

Nicholas was succeeded by his son, Alexander II, who is remembered as the great liberator and reformer. Following Russia's humiliating defeat during the Crimean War, Alexander recognized just how backward Russia was compared to other European powers. In part, he believed that the system of serfdom bore some responsibility for Russia's backward state, and his edict emancipating the serfs was a move towards ending that practice (although the serfs were not really "freed" by this edict). Believing, also, that education was one critical component of the process of modernization, he instituted a series of reforms in education. The University Statute of 1863 reorganized universities and colleges, and provided them with much greater autonomy to run their own affairs. This same statute allowed much greater freedom for students and faculty alike. The emancipation of the serfs had had the consequence of giving rise to a whole new class of citizens, and the illiterate former serfs needed to be educated (at least at some rudimentary level). This was what the Elementary School Statute of 1864 attempted to accomplish by establishing new elementary schools, supported locally, throughout the empire. Later, Alexander extended his educational reforms to include the training of military officers. His Universal Military Training Act of 1874 instituted universal conscription, reorganized the upper echelons of the military, called for improvement in its technology and infrastructure, and established new military schools for the education of officers.

Despite his liberal policies, Alexander was repeatedly the target of assassination attempts. He survived one attempt in 1866, and three more attempts between April 1879 and February 1880. In 1881, however, Alexander fell to a bomb plot. His plans for further reforms, which would have led to an elected parliament and a constitutional monarchy, died with him. His immediate successor, Alexander III, was an anti-reformist, who did much to limit the freedoms that had been granted by his father. Among his acts that are important to this *Brief*, were the Temporary Regulations imposed in 1881, reinforced by the University Statute of 1884, which repealed the autonomy granted to universities by the University Statute of 1863. Nicholas was a confirmed Slavophil, who sought complete Russification of his empire: to homogenize the people, their language, and their loyalty to the Tsar as autocrat. One example of his efforts comes with the Universities of Warsaw and Dorpat, where the language of instruction was changed to Russian from Polish and German.

Alexander III's successor, Nicholas II, did little to alter his father's policies. Thus, he did not repeal the University Statute of 1884, although the universities did regain their autonomy briefly after the Revolution of 1905. Once the Soviet Union came into being, all universities were rigidly and centrally controlled.

1.2 Peter the Great and the Academy of Sciences

Prior to the reign of the modernizing Emperor, Peter the Great, there was no organized secular system of higher education in Russia. Instead, education was in the hands of the clergy, and occurred in the seminaries; the priests who did the teaching were concerned largely with teaching the Russian orthodox faith. Few Russians completed more than an elementary education, and those that did were usually destined for the priesthood. This situation changed in 1725.

Peter's travels in western Europe had revealed just how backward Russia was compared to her rivals in western Europe, and instilled in him the fire to modernize his backward state. So, in 1725, he established the Russian Academy of Sciences, and appointed some of the most distinguished western European scientists as the first Academicians. The first Academy was located in the Kunstkamera (Fig. 1.1), which had been built in 1714 to house the museum he had founded.

Peter did not live to see his vision completed. His policies of modernization were, however, continued by three of the Empresses who succeeded him: Catherine I, Elizabeth, and Catherine II (the Great).

1.3 The Universities and Technological Institutes

Following Peter's establishment of the Academy of Sciences, higher education in Russia advanced only very slowly. It took another three decades before Russia's first secular university was founded—Moscow University, in 1755—and then almost another half century before it was not the sole secular institution of higher education in Russia. At the turn of the nineteenth century, Alexander I began a major decentralization of higher education in Russia with the establishment of new universities at Dorpat (1802), Vilna (1803), Khar'kov and Kazan' (1804). In 1816, Warsaw University (which was closed and reopened multiple times during the nineteenth century) was established, and three years later the Main College at St. Petersburg finally received university status, giving the imperial capital the university that had been part of Peter's vision. Southern Russia received its first university in 1834, with the establishment of St. Vladimir University in Kiev; the second was Novorossiisk Imperial University in Odessa, which was established in 1865. In 1878, the first Russian university in Siberia, the Imperial Tomsk University, was founded to the delight of an enthusiastic local populace.

Fig. 1.1 The Kunstkamera in St. Petersburg, first location of the Imperial Academy of Sciences. © 2006 Sergey Barichev. http://en.wikipedia.org/wiki/File:Kunstkamera_(Saint-Petersburg).jpg

The role of the universities was to carry out advanced study and research, and this role did not change over time. Consequently, there was also a need for practical advanced education, and it was for this purpose that a series of specialized Institutes were founded. Unlike the universities, these colleges focused their attention on the practical education of students for careers in fields such as engineering, architecture and agriculture. In one way, these institutes were actually closer to the vision of Peter the Great for Russian higher education than was his Academy because they focused on technical education instead of the classics. The first of these specialized institutions was the Petersburg Mining Institute, which was founded in 1774, and was one of the first in Europe. The Petrovskaya Academy of Agriculture and Forestry (now the Moscow Timiryazev Agricultural Academy) was founded in Moscow in 1865, and continues as the oldest agricultural college in Russia. The Technological Institutes, which tended to focus on students intending careers in engineering, especially, were founded in university cities, but in most cases, they were not founded until the last quarter of the nineteenth century. The St. Petersburg Technological Institute was founded in 1828, but the Technological Institutes at Moscow (1872), Khar'kov (1885), and Tomsk (1900) were all founded after 1870.

1.3.1 Elizabeth: Moscow University, the First Secular University

Peter's vision had called for the Academy to have an associated university, but it took another quarter century before Russia had its first secular university, and this was founded not in St. Petersburg, but in Moscow (Fig. 1.2), by a 1755 decree of

Fig. 1.2 Moscow University, 1786

Empress Elizabeth [3]. In locating the university in Moscow, Elizabeth was following the suggestion of Academician Mikhail Lomonosov to Count Shuvalov—a court favorite and the Empress' lover—that a university should be founded in the city. Associated with the university were two Gymnasia, one at Moscow, and one at Kazan', on the Volga river 600 miles to the east. This represented the state of higher education in Russia until the turn of the nineteenth century.

Based on Lomonosov's plan, the University was founded with three faculties: Philosophy, where physics and the humanities were taught, Law, where jurisprudence and politics were taught, and Medicine, where chemistry, natural history and anatomy were taught. All students began their studies at the university in the Faculty of Philosophy, where they remained for the first three years, and then they specialized, either continuing in the Faculty of Philosophy or moving into Law or Medicine. The enrolment of the university was open to all persons who could pass the entrance examinations, except serfs.

Chemistry was initially taught in the Faculty of Medicine, but by the turn of the nineteenth century, it had moved into the (new) Faculty of Physical and Mathematical Sciences. Three years after the founding of the university, a small stone building was constructed for use as the chemistry laboratory; the laboratory was renovated periodically over the next century and a half [4].

1.3.2 Paul I: St. Petersburg Medical-Surgical Academy

The Petersburg Medical-Surgical Academy (Fig. 1.3) was founded in 1798 for the purpose of training of military doctors. In 1808, it was raised in rank to "the First Institutions of the Empire", and renamed the Imperial Medical-Surgical Academy. During the nineteenth century, all the experts in surgery and anatomy, and many of those in physiology as well, worked at the Medical-Surgical Academy [5]. The institution was ranked as one of the best educational institutions in Imperial

Fig. 1.3 The St. Petersburg Medical-Surgical Academy (now the Army Medical Academy) in 1914 (photograph by Karl Bulla, 1855–1929)

Russia, and its members enjoyed the rights and privileges of members of the Academy of Sciences (although not the formal admission to membership). The Academy became the Army Medical Academy (its current name) in 1881, and was named after S. M. Kirov from 1935 until the early 1990s, when its current name was restored [6].

1.3.3 Alexander I: New Universities and the First Nationwide, General University Statute

The reign of Alexander I was characterized by extremes, much like the character of the Tsar himself. In the first decade of his reign, Alexander tried to introduce liberal reforms that would ultimately lead to the establishment of a constitutional monarchy. As part of these reforms, he streamlined the government by establishing Ministries, led by ministers directly responsible to the crown, to replace the older Collegia. One of these ministries was the Ministry of Education. In the second half of his reign, however, he became the mirror image of the Tsar who had begun the reign: a reactionary who rescinded many of the liberties he had set in place during the first years of his reign.

As some of his first acts as ruler, Alexander approved charters for the founding or reconstructing four new universities: Dorpat (1803), Vilna (1803), Khar'kov (1804), and Kazan' (1804), along with the University Statute of 1804, the first such statute of general applicability, which established the manner in which the universities were to be governed and operated. Alexander was inspired by enlightened absolutism, and intended to uplift his subjects. In the words of one of his closest advisors, Mikhail Mikhailovich Speransky[2] [7] (Михаил Михайлович Сперанский, 1772–1839), he believed that,

> reforms which are made by the power of the state generally are not lasting. Therefore, it is better and easier to lead people to improvement by simply opening to them the path to their own improvement. Supervising from a distance the peoples' activities, the state can arrange matters to assist them to take the path to improvement without using any kind of force.

Alexander had the choice of following the German or the French model for his new universities. He chose the German model, basing new universities on the University of Göttingen, which had both regional and national missions. Alexander saw the German model as being better fitted to the Russian character, although the differences between Russian and German students meant that some accommodations needed to be made to ensure that the universities ran properly.

1.3.3.1 Dorpat University

Dorpat University (Kaiserliche Universität zu Dorpat, Imperial Dorpat University; now the University of Tartu, in Estonia, Fig. 1.4), had been founded by Gustavus Adolphus II of Sweden in 1632, but it was closed for almost a century (1710–1802) as a result of the Northern War between Russia and Sweden, when all the academics fled to Sweden in advance of the Russian armies [8]. Despite the efforts of the Livonian nobles, and the terms of the treaty of 1710, the university remained closed until it was reopened almost a century later as a German-language, Russian state university by the foundation act approved by Alexander in December, 1802. The university was an important bridge between the Russian empire and Germany, and maintained an essentially German character until 1880, when the process of Russification began. Between 1880 and 1898, the language of instruction, which was German until 1893, and the citizenship of the professoriate were gradually homogenized to Russian; in 1898, the city and the university were both renamed Yuriev, reflecting the completion of the Russification process.

As we will see in later chapters in this Brief, Dorpat University was responsible for the education of some major figures in Russian chemistry, although its impact on organic chemistry was rather less than on physical chemistry and pharmacy. From 1828 to 1838, the university was responsible for the education of many

---

[2] Speransky was the son of a village priest, and educated at the St. Petersburg ecclesiastical seminary, where he rose to become Professor of Mathematics and Physics. Speransky was soon seconded into imperial service, and by 1808 had become an important advisor to Tsar Alexander I. He was a clear-headed, liberal thinker, and his plans for reform in Russia were far-reaching. Despite this, he fell out of favor with the Tsar in 1812, on the eve of war with Napoleon. He was restored to the imperial service in 1816 under Nicholas I, Alexander's successor, and was made Count shortly before his death, in 1839.

Fig. 1.4 Dorpat University in 1860 (Lithograph *Die Universitätsgebäude* by Louis Höflinger, 1860)

future professors in Russian universities through its Professors' Institute. Universities throughout the empire were asked to nominate students with the potential to become professors, and they were educated at Dorpat at state expense, then sent abroad to complete their education, in return for s period of service as professors in Russian universities. Financially, Dorpat was one of the better-endowed universities in the Russian empire during the nineteenth century, and was the staging point for Russian students to continue their advanced studies in Russia and abroad.

1.3.3.2   Vilna University

The Imperial University of Vilna (Fig. 1.5) had begun its existence as the Vilna Academy, founded by Jesuits in 1586 at the request of the Lithuanian nobility [9]. Eleven years later, it was granted the status of a university, and continued its growth for another two centuries. In 1773, the operation of the university was taken over by the Ministry of Education, and it was turned into a modern secular university. Following the Polish-Lithuanian Commonwealth, the university came under the purview of Russia. It received its name as the Imperial University of Vilna in 1803, as a result of Alexander I's statute. The language of instruction, however, remained Polish until the university's closure by Nicholas I, following the 1832 November Uprising, in which the Poles rebelled against Russian rule. The rebellion was put down, and all instruction in Lithuanian and Polish was banned.

Fig. 1.5 The Grand
Courtyard of Vilna
University and the Church of
St. John. ("Album Wileński"
J. K. Wilczińskiego, 1850)

1.3.3.3 Khar'kov University

Khar'kov (modern Kharkiv) received its university (Fig. 1.6) in 1804, largely as a result of the efforts of Vasil Nazarovich Karazin[3] (Ukrainian Василь Назарович Каразін, Russian Василий Назарович Каразин, 1773–1842) [10]. Karazin petitioned the tsar on a number of subjects, but education was close to his heart. So, he urged the tsar to establish the University for Southern Russia in Khar'kov. In 1802, he traveled to the city and gave a speech to the nobility on the benefits of a university, seeking private donations to help establish it. He received pledges to the total of 100,000 rubles, with a pledge to raise more. On his return to St. Petersburg, however, he found that the bureaucracy was unwilling to release the money pledged by the tsar. Impatient, he returned to Khar'kov and spent much

---

[3] Karazin was born into the nobility, and received a good education, including military training in the prestigious Semyonovskii Regiment (Семёновский лейб-гвардии полк), and at the school of mines in St. Petersburg, which had been founded by an edict of Catherine II at the urging of Lomonosov, and had become a premier educational institution in the imperial capital. In his position in the Ministry of Education, Karazin was instrumental in establishing Khar'kov University, but he lost his position before the university was opened. Karazin was forced to return to his village, but did not give up on education: he founded a school for local children. Karazin was a Ukrainian nationalist, and was a strong proponent of a constitutional monarchy; his political views led to his arrest on more than one occasion. The university that he worked to hard to establish is know named V.N. Karazin Kharkiv National University.

Fig. 1.6 Khar'kov Imperial University, prior to 1918

of his own fortune to see the university opened. By the time that the Khar'kov Imperial University officially opened in 1805, Karazin had been dismissed from his post in the ministry, so he did not attend the ceremony. Despite its rocky beginning, by its silver jubilee, the university was firmly established.

1.3.3.4 Kazan' University

When Kazan' was awarded a university (Fig. 1.7) in 1804 as a result of the 1803 charter approved by Tsar Aleksandr I, the response of the local population was tepid, at best. Kazan' had received one of the first two Gymnasia in Russia during the reign of Empress Elizabeth [11], and since the Gymnasium was already affiliated with Moscow University, many of the local population saw no need for a university at Kazan' as well. But… a university was decreed, so a university was established. Teaching at the university was divided into four faculties: Physics–Mathematics, Law, Medicine, and History–Philosophy.

As the economic center of the Volga and Kama River basins, Kazan' was effectively the easternmost outpost of European Russia. In the eyes of most Russians at the time, however, Kazan' was not really a European city, but an Asiatic one, and a less desirable and less prestigious location than any of the other three cities where new universities were to be located. This is one of the reasons why the faculty roster was filled only slowly after its founding: for over a decade the university was only half-staffed, functioning more as an extension of the Kazan' Gymnasium than as an independent institution of higher learning. The other was the mismanagement of the new university by the University Curator (S. Ya. Rumovskii), who was based in St Petersburg, and the Kazan' Gymnasium

Fig. 1.7 Kazan Imperial University ca 1815

Director (I. F. Yakovkin), who supplied him with reports that stated that the
university should come under the direction of the Gymnasium, rather than vice
versa. The official opening of the university occurred on July 5, 1814—over a
decade after its founding [12]. Despite these difficulties, Kazan' University rose to
become one of the preeminent educational institutions in Russia, a position it
retains to the present time [13].

1.3.3.5 St. Petersburg University

In 1780, Catherine had met Emperor Joseph II of Austria, and learned about the
system of schools that his mother, Empress Maria Theresa had introduced five
years earlier. On her return to Russia, she resolved to put the Austrian model into
effect in Russia. In addition to establishing three levels of schools (primary,
middle, and high school), the plan also called for the training of teachers who
could write textbooks, and translate textbooks from western Europe. Initially,
training of teachers was concentrated in the St. Petersburg high school, but in 1786
it was transferred to the Teachers' Seminary. The Teachers' Seminary was closed
in 1801, but was revived in 1803 as the Petersburg Pedagogical Institute, renamed
the Main Pedagogical Institute in 1814. The Main Pedagogical Institute was
located in the Twelve Collegia building (Fig. 1.8), built by Peter the Great to
house the twelve departments of his government. In 1819, Alexander I reorganized
the Main Institute, and converted it into a university consisting of three faculties:
the Faculty of Philosophy and Law, the Faculty of History and Philology, and the

Fig. 1.8   The Twelve Collegia building as shown in a 1753 lithograph

Faculty of Physics and Mathematics. In 1821, its name was officially changed to St. Petersburg Imperial University, and three years later it received its official charter, which was modeled on the charter of Moscow University [14].

1.3.3.6   Warsaw University

The history of Poland is a tortured one. Poland as an independent kingdom was gradually eliminated by the Partitions of the Polish-Lithuanian Commonwealth, which divided the lands of the kingdom among the Russian empire, the Kingdom of Prussia, and Habsburg Austria; by 1795, Poland-Lithuania had ceased to exist. At the end of the Napoleonic wars, the Kingdom of Poland (also known as Congress Poland) was established from the Duchy of Warsaw as a nominally independent kingdom. However, the new kingdom was ruled by the Russian tsar, so the union of Poland and Russia was a formal fact, at least. In 1816, one year after the establishment of the kingdom by the Congress of Vienna in 1815, the charter for the Royal University of Warsaw (Fig. 1.9) was approved; the university was initially composed of the departments of Law and Administration, Medicine, Philosophy, Theology, and Art and Humanities [15]. Fourteen years later, the university was closed as a result of the November 1830 Uprising. The university reopened in 1857, after the Crimean war, as a Medical-Surgical Academy. In 1862, departments of Law and Administration, Philology and History, and Mathematics and Physics were opened. Within a short time, the Academy had been renamed the

Fig. 1.9 Main gate to the University of Warsaw. A view from the Institute of Philosophy, May 2007

"Main School." The Main School was short-lived; following the 1863 Uprising, all Polish-language institutions were closed; in 1869, the Main School was itself closed, and replaced in 1870 by the Russian-language, Imperial University of Warsaw.

1.3.4 Nicholas I: Reaction

Just as the early years of the reign of Alexander I are characterized by a liberal bias, the entire reign of Nicholas I must be described as reactionary. It was Nicholas who closed Vilna University, where instruction had been carried out in Polish, and Nicholas who decreed that education in the empire would be conducted in Russian (only Dorpat University escaped this decree). Although Nicholas was a martinet in much of what he did, he did correct many of the excesses of the previous reign. He appointed Uvarov as Minister of Education; despite his "Orthodoxy, Autocracy and Nationality" program, Uvarov was genuinely interested in promoting learning, and was generally a positive influence on the development of higher education in Russia. Equally importantly, Nicholas removed the odious Magnitskii and his peers from their curatorships of the universities.

Fig. 1.10   St. Vladimir University, Kiev, early twentieth century (before 1918)

1.3.4.1   St. Vladimir University of Kiev

There was just one university founded during the reign of Nicholas I: Kiev University (now Taras Shevchenko National University of Kyiv [16], Fig. 1.10), which was founded in 1834. The university was founded as St. Vladimir University shortly after Vilna University had been closed, and it benefited from the transfer of assets from that institution. Originally, it was composed of a single Faculty: the Faculty of Philosophy, which was divided into the Department of History and Philology and the Department of Physics and Mathematics. A year later, the Faculty of Law was added, and in 1847, the Faculty of Medicine. Later, the two departments of the original Faculty of Philosophy became independent Faculties of the university.

1.3.4.2   St. Petersburg Technological Institute

One of Peter's motivations in founding the Academy of Science had been to promote technical education in Russia. A century after the founding of the Academy of Sciences, Nicholas I authorized the founding of the Petersburg Practical Technology Institute—the first such institute in the empire—as an institution dedicated to technical education [17]. The subjects taught are mainly characterized by being highly applied (e.g. architecture, civil engineering, practical mechanics, mechanical technology, chemical technology, metallurgy,

Fig. 1.11 St. Petersburg Technological Institute, ca. 1829

drawing, and designing were all subjects taught at the Institute. In 1895, the Institute (Fig. 1.11) was renamed the Petersburg Technological Institute, and named for Nicholas I.

The Institute began as a boarding school, with the boarders paying fees. The fees were waived for students studying to become engineers and manufacturing specialists. Alhough both Mendeleev and Beilstein served on the faculty during the last half of the nineteenth century, the Institute was not designed to be a research-intensive institution, and the research output from the institute was minimal. Nevertheless, three of its faculty members (Mendeleev, Beilstien, and Shulyachenko) were founding members of the Russian Physical–Chemical Society in 1868.

1.3.5 Alexander II: University Expansion Continues

Nicholas was followed to the throne in 1855 by Alexander II, the great reformer and emancipator. In 1855, he repealed the limitations on the number of university admissions, but, at the same time, he decreed new regulations on entrance examinations and fees. This angered the students, who were even more inflamed by the government's refusal to permit the formation of independent student corporations within the universities; in 1861 there were student riots at Kazan' and St. Petersburg. To some degree, the University Statute of 1863 ameliorated the situation, but discontent still simmered below the surface, as evidenced by the numerous asassination attempts on the tsar.

Fig. 1.12 Imperial Novorossiisk University

1.3.5.1 Novorossiisk Imperial University (Odessa University)

Ten years after Alexander's ascention to the throne, he decreed a new university (Fig. 1.12) in Odessa (now in the Ukraine) [18]. In 1803, Armand-Emmanuel de Vignerot du Plessis, Duc de Richelieu, a French Rroyalist who had left Paris at the order of Marie Antoinette before the Reign of Terror, was appointed by Tsar Alexander I as governor of the city; he retained that role until his return to France in 1814. What became Odessa University began its existence as the Lycée Richelieu, which had been founded in 1817. In 1855, Mendeleev had taught here while working on his M. Chem. dissertation. In 1865, the Lycée formed the basis for the Imperial Novorossiisk University, which began with three departments: History and Philology, Physics and Mathematics, and Law; Medicine was added in 1900. Chemistry at Odessa did not have especially auspicious beginnings, but it soon attracted some eminent chemists: in 1872, Markovnikov was called to the Chair of Chemistry there, and Nikolai Dmitrievich Zelinskii (Николай Дмитриевич Зелинский, 1861–1953), Markovnikov's successor at Moscow, received his education there.

1.3.5.2 The Imperial Siberian University (Tomsk)

In 1878, over a decade after the foundation of the Imperial Novorossiisk University, Alexander decreed a new university (Fig. 1.13) in Siberia, in the city of Tomsk [19]. This was the first university east of the Urals, and the only university between the Urals and the Pacific. The idea of opening a university in Siberia had occurred to progressives in Russia as early as 1803, but it was not for another three quarters of a century that it became reality. The government kept delaying the

Fig. 1.13 The Imperial Siberian University in Tomsk, 1900

Fig. 1.14 Riga
Polytechnicum, 1862–1896

decision due to a lack of money or the inadequate development of secondary education in the region. Moreover, some thought that a university in Siberia was a luxury—and it was dangerous to give Siberians a higher education. However, by the end of the nineteenth century, it became obvious that these delaying tactics had run their course.

The construction of the main building and the first dormitory of Tomsk Imperial University were made possible by private donations, which in fact, amounted to half of the budget. It is important to note that while this construction was under way, a library was being assembled, as well as the materials for laboratories, museums, the botanical garden with its greenhouse, and the herbarium. From the beginning, the university was enthusiastically supported by the local population.

1.3.5.3 Riga Polytechnicum

During the reign of Alexander, the first Polytechnicum in the Russian empire— Riga Polytechnicum, in Latvia (Fig. 1.14)—was founded by the Baltic nobility, and modeled on the Higher Technical Schools of western Europe [20]. The charter for the Polytechnic was approved by the tsar in 1861, and the Polytechnicum began accepting students into its Preparatory School in 1862. In 1863, the Polytechnicum was composed of four departments: Engineering, Chemistry, Agriculture, and Mechanics; departments of Commerce and Architecture were added in 1868 and 1869, respectively.

This institution differed from the universities in the Russian empire several important respects: First, it was a private institution. Then, unlike Russian universities, which had begun permitting education of women, enrolment was restricted to males until 1917. However, entry was open to all men who could pay the fees, regardless of nationality, religion, or social status; there was no entrance examination. Instead, students began in the associated Preparatory School, since many of the students seeking admission were not well prepared for tertiary education. The Preparatory School was finally closed in 1892. Riga Polytechnicum was also the first multi-branch institution of higher learning in the Russian empire. Like the other university founded by Baltic Germans, Dorpat, the language of instruction was German until 1896, when the Russification of universities in the empire was begun in earnest by Nicholas II.

References

1. Russian uses the Cyrillic alphabet, and so names must be transliterated to the Roman alphabet. The exact transliteration used depends on the language into which the transliteration occurs, and even this is not a constant within the same language. A good example of this is provided by the name of N. N. Sokolov, which is transliterated into German as Socoloff by *Justus Liebigs Annalen der Chemie*, and as Sokoloff by *Erdmans Journal für Praktische Chemie*. Throughout this book, the BGN/PCGN romanization system for Russian is used as the most intuitive for English speakers. The consonants ж, ц, ч, ш, and щ are transliterated as zh, ts, ch, sh, and shch, respectively. The vowels й, ы, э, ю, and я are transliterated as i, y, e, yu, and ya, respectively. The vowel e at the beginnings of words is transliterated as ye. The soft sign (ь) is rendered as ', and the hard sign (ъ) is rendered as ". In 1918, the Soviets consolidated the orthography of Russian alphabet, eliminating three letters, and dramatically reducing the use of the hard sign at the end of words; the Russian spelling of titles of articles has been modernized prior to transliteration. In citations of articles in western journals, names are given as transliterated by the journals.
2. Hosking G (1998). Russia: People and Empire 1552-1917. Harvard University Press, Cambridge, p. 146.
3. It is one of the joys for anyone involved in historical research about Russian universities that most have excellent official web sites. It is the author's intention to cite these web sites throughout this chapter. The web site for the history of Lomonosov Moscow State University can be found at: http://www.msu.ru/en/ [accessed October 12, 2011].
4. Evans CT (1991). Count Sergei Stroganov and the Development of Moscow University, 1835-1847. PhD Diss, University of Virginia.
5. Luyendijk-Elshout AM Medicine. In Rüegg W (2004). Universities in the nineteenth andearly twentieth centuries (1800-1945). Cambridge University Press, Cambridge. Ch 14, pp 543-590.

6. Considerable information about this and other institutions of higher education in St. Petersburg can be found at the site for the St Petersburg Encyclopedia: http://www.encspb.ru/en/ [accessed October 12, 2011]. Information specific to the Medical-Surgical Academy can be found at http://www.encspb.ru/en/article.php?kod=2804015556 [accessed October 12, 2011], and at http://ru.wikipedia.org/wiki/Военно-медицинская_академия_им._С._М._Кирова (in Russian).

7. (a) Cited by Flynn JT (1988). The University Reform of Tsar Alexander I, 1802-1835. Washington: The Catholic University of America Press, p. ix. (b) Dohnt P (2008) Ambiguous Loyalty to the Russian Tsar. The Universities of Dorpat and Helsinki as Nation Building Institutions. Hist. Social Res. 33:99-126.

8. The history of Tartu University, written by Sirje Tamul, of the Department of History, may be found at: http://www.ut.ee/en/university/general/history [accesses November 10, 2011]

9. For a brief history of the university at its official website, see: http://www.vu.lt/en/about-us/history [acessed November 10, 2011]

10. An official history of the university can be found at its website, managed by Maksim A. Folomieiev: http://www.univer.kharkov.ua/en/general/our_university/history [accessed November 10, 2011]

11. Seton-Watson H (1967). The Russian Empire, 1801-1917. Oxford University Press, Oxford, p. 35-36.

12. Vinogradov SN (1965). Chemistry at Kazan University in the nineteenth century: A case history of intellectual lineage. Isis 56:168-173.

13. The university has an excellent website. For the general history of the university, see: http://www.ksu.ru/eng/general/history.htm [accessed November 10, 2011]. The museum of the History of Kazan University (S. V. Pisareva, Director) has an official website at: http://www.ksu.ru/miku/eng/index.htm [accessed November 10, 2011]. The chemical museum at Kazan (Kand. T. D. Sorokhina, Director) also has an excellent website: http://www.ksu.ru/chmku/eng/ [accessed November 10, 2011].

14. It is more difficult to find information about the history of St. Petersburg State University from its website, http://eng.spbu.ru/ [accessed November 10, 2011; it is suggested that one search the site under "history," which brings up a number of hits, including the sites of several museums.

15. The index page for the history of Warsaw University can be found on its official website at: http://www.uw.edu.pl/en/page.php/about_uw/history.html [accessed November 10, 2011]. This site has links to several pages discussing the university through history.

16. The history of the university can be found at the official website, http://www.univ.kiev.ua/en/geninf/history [accessed November 10, 2011]

17. The history of the Institute can be found at its official website, at: http://www.spbtechnological university.com/Lists/History%20of%20the%20Institute/Foundation%20and%20Formation%20of%20the%20Institute%20as%20an%20Institution%20of%20Higher%20Education.aspx [accessed November 10, 2011].

18. The history of Odessa national University named for I. L. Mechnikov is found at: http://onu.edu.ua/en/geninfo/history [accessed November 10, 2011].

19. The history of Tomsk State University can be found on its official website at: http://www.tsu.ru/WebDesign/TSU/coreen.nsf/structurl/history_doc1 [accessed November 10, 2011]

20. The history of the polytechnicum can be found at the official website of Riga Technical University: http://www.rtu.lv/en/content/view/1464/1168/lang,en/ [accessed November 10, 2011].

Chapter 2
Beginnings

2.1 Introduction

At the start of the twentieth century, organic chemistry was not yet 75 years old as a separate and legitimate sub-discipline of the science. Considerable progress had been made in these first seven decades, and the stage was set for the dramatic advances in the science to come in the following century. Most practicing organic chemists are familiar with many of the great German, French and English organic chemists whose work helped the fledgling discipline grow, but few are familiar with the role that Russian organic chemists of the nineteenth and early twentieth century played in the development of the science. And this is in spite of the fact that many of the named rules and reactions that one studies in the first course in organic chemistry are, in fact, of Russian origin. It is the intent of this book to help rectify that deficiency.

2.2 The Early History of Organic Chemistry

As a separate discipline, most organic chemists trace the origins of their science as a separate sub-discipline of chemistry to the serendipitous synthesis of urea by Friedrich Wöhler (1800–1882), reported in 1828 [1]. This had not been the first synthesis of urea, which had actually been accomplished by John Davy (1790–1868), in 1811 [2], and by Wöhler himself in 1824 [3], but Wöhler did something that Davy did not—he recognized that he had *not* prepared the ammonium cyanate he was trying to make. After four years of careful work, he succeeded in identifying the water-soluble non-electrolyte that he had prepared as urea.

Friedrich Wöhler (1800–1882) had entered university at Heidelberg, where he underwent the training required to become an obstetrician. His heart, however, had always belonged to chemistry, and on the advice of Leopold Gmelin (1788–1853),

D. E. Lewis, *Early Russian Organic Chemists and Their Legacy*,
SpringerBriefs in History of Chemistry, DOI: 10.1007/978-3-642-28219-5_2,
© The Author(s) 2012

he forsook medicine and pursued a chemical career instead, moving to Sweden to study with the great Swedish chemist, Jöns Jacob Berzelius (1779–1848). It was in Berzelius' laboratory that he began the study of cyanates that would end with his synthesis of urea. At heart, however, Wöhler remained an inorganic chemist, despite his lifelong friendship and collaboration with Justus Liebig (1803–1873), whom he had met as a young man. Among his accomplishments are the isolation of aluminum, beryllium, and yttrium in metallic form.

Friedrich Wöhler (1800-1882)

Jöns Jacob Berzelius (1779-1848)

Many introductory textbooks of organic chemistry imply that this synthesis of urea—as Wöhler told his mentor, Berzelius, "without the need for a kidney"—spelled the end of vital force theory, only just formalized by Jöns Jacob Berzelius in the second decade of the nineteenth century [4]. Of course, this is not the case. Chemists, like other scientists, are reluctant to abandon the old, familiar theories that form the framework of their science, preferring to modify them until new experimental evidence finally makes the theory totally untenable. Urea, for example, could be argued as being an excretion product, and therefore devoid of its vital force. Thus, it took another three decades, the synthesis of acetic acid from its elements by Hermann Kolbe (1818–1884) in 1845 [5] and the publication of Marcelin Berthelot's *Chimie organique fondée sur la Synthèse* in 1860 [6] before vital force theory was finally consigned to the scrapheap of organic chemistry. Kolbe's synthesis was the first in the modern meaning of the language: In the reaction sequence he used, acetic acid was prepared from elemental carbon, sulfur, chlorine, and water, by a carefully designed sequence of reactions. More importantly, none of the starting materials in this sequence was ever associated with having a vital force.

The rapidity with which vital force theory had attracted adherents is partly due to its first major proponent. Jacob Berzelius was, at the time, the most influential chemist in Europe: if he accepted an idea, it was accepted universally, if he did not, it was certain to be consigned to obscurity. Some evidence of his impact can be gauged even today: the terms, catalysis, polymer, allotrope, halogen, and protein, for example, were all suggested by Berzelius. It helped, of course, that he had been responsible (in part, at least) for the training of many of Europe's leading young chemists. Following his graduation from the Gymnasium, Berzelius studied medicine at the University of Uppsala, graduating with his M.D. in 1804. However, he soon turned his efforts to chemistry. He was one of the first to embrace Dalton's atomic theory, and spent a considerable part of his early professional life to gaining evidence to support the Law of Definite Proportions; in the course of this work, he developed the first accurate table of atomic weights. He also developed the modern chemist's short-hand for writing formulas, replacing John Dalton's pictograms by the modern one- or two-letter symbols for the elements. From his researches with the electricity, he developed the theory that atoms formed certain stable groupings, which he termed, "radicles," that behaved as a single unit in chemical reactions. He discovered selenium (1818), silicon (1824) and thorium (1829). On the occasion of his marriage, in 1835, Berzelius was created baron by King Charles XIV of Sweden.

2.3   The Western European Schools

The three decades following Wöhler's serendipitous discovery of the conversion of ammonium cyanate to urea were dominated by contributions from France and, more especially, Germany. The demise of vital force theory coincided to some degree with the rise of the great German and French schools under Liebig, Wöhler, Kolbe, Robert Bunsen (1811–1899), and Emil Erlenmeyer (1825–1909) in Germany, and Adolphe

Wurtz (1817–1884) in Paris. Somewhat later, the English schools established by August Wilhelm von Hofmann (1818–1892) and Alexander William Williamson (1824–1904) rose to a level of prominence to join their continental counterparts. During this period, the concepts of radicals, types, and substitution all came into their own, until they were supplanted in 1858 by the structural theory of organic chemistry [7]. During this time period, progress in organic chemistry was hampered by the lack of a uniform set of atomic weights, with many chemists rigidly adhering to the equivalent weights of elements ($C = 6, O = 8$). This practice led to the necessity of using "double atoms," often written with a barred symbol, in order to write acceptable formulae for organic compounds. The values for the atomic weights of the elements were finally settled by the Italian chemist, Stanislao Cannizzarro (1826–1910), at the Karlsruhe conference of 1860; the preceding decades had seen a gradual separation of meaning for the terms, "molecule," and "atom." The concept of valence, which was critical for the development of the structural theory of organic chemistry, was proposed by English chemist, (later Sir) Edward Frankland (1825–1899), in 1852, following his studies with the dialkylzincs.

The underpinning of each of these theories was due to observations by a series of organic chemists. The fruitful collaboration between Liebig and Wöhler resulted in the seminal study of benzoyl compounds that led to the concept of the polyatomic radical [8]. Liebig further clarified his definition of a radical in 1837, where he stated [9]:

> We call cyanogen a radical (1) because it is a non-varying constituent in a series of compounds, (2) because in these latter it can be replaced by other simple substances, and (3) because in its compounds with a simple substance, the latter can be turned out and replaced by equivalents of other simple substances.

Adolph Wilhelm Hermann Kolbe (1818-1884)

Sir Edward Frankland (1825-1899)

Justus von Liebig (1803-1873)

Charles Adolphe Wurtz (1817-1884)

Robert Wilhelm Eberhard Bunsen
(1811-1899)

Richard August Carl Emil Erlenmeyer
(1825-1909)

August Wilhelm von Hofmann
(1818-1892)

Alexander William Williamson
(1824-1904)

2.4 The Origins of Russian Science

The year 1725 marks a watershed in the development of science in Russia. In that year, Peter the Great established the Russian Academy of Sciences, and appointed some of the most distinguished western European scientists as the first Academicians. For the next seventeen years, until 1742, all appointments to the Academy were from western Europe, but in 1742 that situation changed with the appointment of Mikhail Vasil'evich Lomonosov (Михаил Васильевич Ломоносов, 1711–1765) as an Adjutant Member of the Academy. Lomonosov was elected a Full Member of the Academy three years later.

2.4.1 Lomonosov and the Early Academy of Sciences

A true renaissance man and an obvious genius, Lomonosov not only contributed to the development of science in Russia, but he made major contributions to Russian language and poetry. With his patron, the Count Ivan Ivanovich Shuvalov (Иван Иванович Шувалов, 1727–1797),[1] he founded Moscow University (now Lomonosov Moscow State University).

---

[1] Shuvalov was a leader of the enlightenment in Russia following the death of Peter the Great. A favorite (and lover) of Empress Elizaveta Petrovna (r. 1741–1762), he was the first Russian Minister of Education, and was instrumental in establishing Moscow University, of which he became the first Curator. He was a strong patron of the arts, and Russia's first theater and its first Academy of Arts were established at his initiative; he served as President of the Academy of Arts from 1857–1862. The Gymnasia at St. Petersburg and Kazan' were also the results of his efforts.

In many ways, Lomonosov was a man ahead of his time. He was a courtier, a poet, and a scientist of the first rank [10]. Three decades before Lavoisier wrote his treatise against phlogiston [11], Lomonosov came to the conclusion that the theory of phlogiston was incorrect, and showed, in an unpublished memoir in 1756, that the oxidation of a metal in a hermetically sealed container leads to no change in mass [12]. In 1673, Robert Boyle had carried out the same experiment, calcining lead in a sealed retort. However, he had opened the vessel in which the metal was fired before weighing it, and found that the weight of the calcination product exceeded the weight of the original metal. From this observation, he concluded that "corpuscles of fire" had passed through the glass of the retort and were absorbed (fixed) by the metal [13]. Lomonosov's observations that there was, in fact, no increase in weight until air was admitted to the vessel, led to his proposal of a form of the Law of Conservation of Matter [14]:

> Every change that takes place in nature occurs in such a way that if something is added to something else, the same is subtracted from another body. Thus matter added to one body is lost by another. The number of hours I sleep is subtracted from the time I am awake, and so on. Since this is a universal law of nature, it also governs the rules of motion: a body which jolts another body to move loses as much of its motion as it imparts to the one it started moving.

Михаил Васильевич Ломоносов
(1727-1797)
(Mikhail Vasil'evich Lomonosov)

During the harsh winter of 1759, he and Academician I. A. Braun were the first to observe the freezing of mercury and to describe the properties of the solid metal. Lomonosov made contributions to the theory of heat (which he believed to be associated with motion); as a result of these studies, he promulgated his "universal law," a law of conservation of matter and energy [15], and the wave theory of light, and he first alluded to the principle that would later be known as the conservation of matter.

The rise of chemistry in Russia, and of organic chemistry later, required an infrastructure that was not present in 1725, when Peter the Great founded the Academy of Sciences in St. Petersburg. Due to the lack of qualified Russians, Peter populated his new Academy with eminent foreign scientists. Over the next quarter century, much of the progress in chemistry in Russia was due to an influx of (especially German) chemists from abroad, as Peter's successors opened Russia to foreign scientists. However, by the middle of the century, that had begun to change. By that time, Lomonosov, the first Russian Academician, had risen to a position of some power in the Academy, and he was constantly at loggerheads with the leaders of the German faction in the Academy, who saw his rise as a threat to their continued dominance of the Academy.

2.4.2 Russification of Russian Chemistry

In the mid 1830s, there was a general movement towards nationalism throughout Europe, and Russia was no exception to this trend. The nationalist movement in Russia was embraced by Tsar Nicholas I, whose reign had begun in 1825 with what is known as the Decembrist revolt. This failed revolt had the long-term effect that Nicholas and his advisors had a distrust of foreigners, whom they suspected of planning to foment revolution (unreasonable, perhaps, but not baseless, given the political turmoil in western Europe at the time). Consequently, around 1835, a movement towards Russification of the universities in Russia began in earnest. As part of this process, severe limits were placed on any activity that might be described as "freethinking," and there was an attempt to forestall the entry of foreign (read, "revolutionary") educators into Russia. However, Nicholas' attempt to promote Russification by establishing a purely Russian "Professional Institute" failed to meet the need for qualified teachers. This meant that Russian students still needed to go abroad to western Europe to complete their technical education; students abroad—especially at progressive institutions such as Giessen—were closely monitored by the Tsar's secret police.

Up to this time, Chemistry Departments had been staffed by a preponderance of foreign (usually German) Professors, partly because of a lack of suitably qualified Russian candidates, but also partly because of a perceived inferiority in those Russian candidates. As the Russification process gained momentum, more and more well-trained young Russian scientists began to be appointed to Chairs of Chemistry at Russian universities, and science began to lose its "foreign" trappings. But one should not infer from this that Russification proceeded smoothly or uniformly. It did not. In fact, the Academy of Science remained solidly in foreign hands, and responded to Russification only slowly (much more slowly than the universities), so that the most prestigious positions were still occupied by non-

Russians. Of course, this situation could not last, and as the nineteenth century progressed, a gradual split appeared in the Russian Academy of Sciences, with members in the "Russian" party, and other members in the "German" party. This split came to a head in 1880, when Dmitrii Ivanovich Mendeleev (Дмитрий Иванович Менделеев, 1834–1907) was denied the chair in technology of the Academy of Sciences in St. Petersburg by a single vote [16]. Two years later, the position went to Fyodor Fyodorovich Beil'shtein (Фёдор Фёдорович Бейльштейн, Friedrich Konrad Beilstein, 1838–1906) who was viewed by the Russian party as a German, despite having been born in St. Petersburg, and having taken the unusual step of being naturalized a Russian citizen [17]. Nevertheless, by the turn of the twentieth century, chemistry in Russia was truly Russian, with that generation of organic chemists being able to point to significant contributions by earlier generations of their countrymen.

Дмитрий Иванович Менделеев (1834-1907)
(Dmitrii Ivanovich Mendeleev)

Фёдор Фёдорович Бейльштейн (1838-1906)
(Friedrich Konrad [Fyodr Fyodorovich] Beilstein)

One of the major problems that one encounters when studying the history of organic chemistry in Russia comes from the politicization of science and the history of science during the Soviet era. One particularly illuminating example of this concerns the structural theory of organic chemistry, and the part played in its development by Aleksandr Mikhailovich Butlerov (Александр Михайлович Бутлеров, 1828–1886). In 1861, Butlerov had made a presentation at the Chemistry Section of the 36th Congress of German Physicians and Scientists. His presentation "Einiges über die chemische Structur der Körper," was published in the *Zeitschrift für Chemie und Pharmacie* [18]. In that paper, Butlerov introduced the term, "chemical structure," and made the point that each compound had one, and only one, structure.

In the west, Butlerov's contributions were largely overlooked until the last quarter of the twentieth century. In the Soviet Union, however, his contributions were exaggerated to the point that he was, at times, viewed as the founder of the science of organic chemistry. His contributions were most blatantly politicized during the controversy about the theory of resonance in the Soviet Union in the early 1950s, when what should have been a scientific argument degenerated from science into what amounted to jingoism. As described by Hargittai [19]:

> The critics of the theory of resonance contrasted Butlerov's true Russian values with the cosmopolitan views of those who had bowed slavishly to Western values, etc. The proponents of the theory of resonance had to exercise humiliating self-criticism and lost their jobs [9].[2] The minutes of a meeting in Moscow on June 11–14, 1951, were published in a

440-page hardbound volume [10].[3] Four hundred and fifty chemists, physicists, and philosophers attended the meeting, including the top chemists from all over the Soviet Union. There was a report on "The status of chemical structure theory in organic chemistry" compiled by a special commission of the Chemistry Division of the Soviet Academy of Sciences. It was followed by forty-three oral contributions. The report consisted of eight chapters and the first was titled "Butlerov's teachings and their role in the development of chemistry."

Александр Михайлович Бутлеров (1828-1886)
(Aleksandr Mikhailovich Butlerov)

This need to preserve the pre-eminence of the Russian chemist over western chemists often led to Soviet historians claiming credit for Russian chemists for discoveries where the actual discovery was not, in fact, made by the Russian. What they missed, which is important, is that there are instances where the original idea was not due to the Russian chemist, but the clarification that improved its utility— often a much more important contribution—was. This was certainly the case with Butlerov, whose status as a giant among Russian organic chemists of the nineteenth century needs no artificial enhancement by claims of accomplishments beyond what he actually did. Nevertheless, despite the promulgation of dubious claims of priority, the seminal contributions of Russian organic chemists to the

---

[2] Hargittai I (2000) The great Soviet resonance controversy. In Hargittai M (ed) Candid Science: Conversations with Famous Chemists. Imperial College Press, London, pp 8–13.

[3] Sostoyanie teorii khimicheskogo stroeniya v organicheskoi khimii (The state of affairs of the theory of chemical structure in organic chemistry). Publishing House of the Soviet Academy of Sciences, Moscow, 1952.

development of the discipline cannot be gainsaid. In the chapters that follow, I hope to show just how large a debt modern organic chemists owe to these early Russian pioneers of their science

2.5 Progress Through an Academic Career in Nineteenth-Century Russia

As part of his reforms associated with the establishment of the Academy of Sciences, Peter had envisioned a university associated with the Academy, and a Gymnasium associated with the university. Although Peter died before he could put this plan into action, his successors eventually did just that. The founding of Moscow University was accompanied by founding of two Gymnasia that were directly associated with it: at St. Petersburg, and in the eastern city of Kazan' (Казань), which was to flower into a major center for Russian organic chemistry. The secondary school system in Russia through the time of Nikolai I was heavily stratified, with entry requirements based, in part, at least, on the social class of the student. Serfs and household servants could aspire to an elementary education at parish and district schools, but the Gymnasium was reserved for the children of nobles or officials. The children of priests could obtain their education free at the seminaries, and many did, although the fraction who actually then entered the priesthood was small. The universities, in contrast, were open to all free persons above sixteen years of age who could pass the entrance requirements. Barriers to the education of serfs did not fall appreciably until the Emancipation decrees of Aleksandr II.

Following their graduation from the Gymnasium, students would enter the University by taking entrance examinations (of which Latin was an important component), and then entering an appropriate faculty. The first degree obtained by a student was the *diplom*, which approximates the modern baccalaureate degree. Following the *diplom*, students seeking an academic career would undertake additional advanced study, and would complete a research project on which they would write a dissertation for the degree of *kandidat*. In the nineteenth century, the *kandidat* was approximately at the level of a modern master's degree; today it is the full equivalent of the Ph.D. Completion of the *kandidat* qualified the student for entry-level positions in the university, almost always as Laboratory Assistants attached to the Chair (*kafedra*) of Chemistry.

The next step up the academic career ladder required a more advanced degree, the *Magistr Khimii* (M. Chem.), which was obtained by undertaking a research project, and writing up the results in a dissertation. By the end of the nineteenth century, not only was the dissertation assessed by a committee of examiners, but the candidate was also required to give a public oral presentation on the subject of the dissertation, and to pass an oral examination on the subject of the dissertation. Unlike the *kandidat* dissertation, there was a realistic expectation that the M. Chem. dissertation would contain work at the publishable level, and relatively few individuals obtained the degree without having the work in their dissertations also appear in the scientific literature. Obtaining the degree of M. Chem. permitted the student to obtain an entry-level faculty appointment, typically at the rank of Adjunct (Assistant Professor), and students could progress from there to the rank of *Extraordinarius* (Associate Professor) without further qualification.

The highest-ranked position in a Russian University was the Professor, who occupied a Chair (*kafedra*) in a discipline. The number of such chairs available at the university was fixed, which meant that the number of professors was also fixed. In order to occupy a *kafedra* in chemistry, the degree of *Doktor Khimii* (Dr. Chem.) was required. This degree approximates the higher earned doctorates (e.g. the Doctor of Science) awarded by British Commonwealth universities, or the *habilitation* in the German system. In order to qualify for the degree, an individual was required to apply for permission to present for the degree. On obtaining permission to become a candidate for the degree, the individual would present a dissertation based on his original research; in most cases, the dissertation contained material that had already been published in peer-reviewed journals. In addition, the candidate was required to sit examinations in all areas of chemistry, and to make a public oral presentation on the dissertation. Only on passing all the examinations, having the public presentation assessed as satisfactory, and having the committee of examiners find the dissertation of sufficiently high standard, did the candidate receive the degree. The degree of Dr. Chem. is still the highest earned degree in chemistry conferred by Russian universities [20].

References

1. (a) Wöhler F (1828). Ueber künstliche Bildung des Harnstoffe. Pogg Ann Phys Chem 12:253-256. (b) Wöhler F (1828) Sur la Formation artificielle de l'Urée. Ann Phys Chem 37:330-333.
2. Davy J (1813). On a gaseous compound of carbonic oxide and chlorine. Ann Phil [1] 1:56-57. For earlier reports by Davy, see: Davy J (1811) Nicholson's J. 30:28; (1812) 32:241-247; Davy J Phil Trans Roy Soc, London 104:144.
3. Wöhler F (1824). Om några föreningar af Cyan. Kongl Vetens Akad Handl 328 (in Swedish); (1825). Ueber Cyan-Verbindungen. Pogg Ann Phys Chem [2] 3:177-224 (in German).
4. (a) Berzelius J (1814). Experiments to determine the definite proportions in which the elements of organic nature are combined. Ann Phil 4:323-331, 4:401-409; (1815). 5:93-101, 5:174-184, 5:260-275. This work also appeared in Swedish: Berzelius J (1818). Försok öfver de bestamda förhållanden, hvari elementen för den organiska naturen äto förenade. Afhandlingar 5:520-646. The specific description of vital force is in the first paper, p. 328. (b) Berzelius J (1827). Lehrbuch der Chemie. 1st edn. Arnoldischen Buchhandlung, Dresden and Leipzig. Vol. 3, pp. 135-138.
5. Kolbe H (1845). Beiträge zur Kenntniss der gepaarten Verbindungen. Ann Chem Pharm 54:145-188.
6. Berthelot M (1860). Chimie organique fondée sur la Synthèse. Mallet-Bachelier: Paris.
7. For a recent account of the development of organic structural theory, see: Lewis DE (2010). 150 Years of organic structures. In Giunta C.J (ed) Atoms in Chemistry: From Dalton's Predecessors to Complex Atoms and Beyond. ACS Symp Ser 1044:35-57.
8. Wöhler F, Liebig J (1832). Untersuchungen über das Radikal der Benzoesäure. Ann Chem Pharm 3:249-287.

9. Cited in: Ladenburg A (1900) Lectures on the History of the Development of Chemistry Since the Time of Lavoisier. Dobbin L (translator). Alembic Club, London.

10. For reviews of Lomonosov's life and work, see: (a) Menshutkin BN (1952). Russia's Lomonosov, Chemist Courtier, Physicist Poet. Princeton University Press: Princeton. (b) Kuznetsov BG (1945). Lomonosov, Lobachevskii, Mendeleev. Academy of Sciences Press: Moscow and Leningrad. (c) Leicester HM (1961). Vladimir Vasil'evich Lomonosov. In Faber E (ed.) Great Chemists. Interscience, New York.

11. (a) Lavoisier AL (1783). Réflexions sur le phlogistique. Mém Acad Sci, 505-538. Reprinted (1783) as Réflexions sur le phlogistique, pour servir de développement à la théorie de la combustion & de la calcination, publiée en 1777. Académie des sciences: Paris. (b) This paper was the first of a series: Duveen, DI, Klickstein HS (1954). A Bibliography of the Works of Antoine Laurent Lavoisier. Dawson and Sons and E. Weil: London., papers 35, 38-43, 45 and 49.

12. Lomonosov V (1747-1748). Novi Commentarii Academiae Scientiarum Imperialis Petropolitanae 1:230-244.

13. Boyle R (1673). Essay of the Strange Subtilty, Great Efficacy, Determinate Nature of Effluviums. To whieh are annext New Experiments to make Fire and Flame Ponderable: Together with a Discovery of the Perviousness of Glass. Printed by W. G. for M. Pitt: London.

14. (a) Pavlova GE, Fedorov AS (1984). Mikhail Vasilievich Lomonosov: His Life and Work. Aksenov A (transl.) Mir, Moscow; distributed in the U.S.A. by Imported Publications, Chicago; pp 151-152. (b) Kauffman GB, Miller FA (1988). Mikhail Vasil'evich Lomonosov (1711-1765): Founder of Russian science, a philatelic portrait. J Chem Educ 65:953-958.

15. Lomonosov V (1760). Meditationes de Solido et Fluido (Contemplations on Solid and Liquid Bodies). Presented on September 6, 1760.

16. For an account of this vote and its aftermath, see: Gordin MD (2004). A Well-Ordered Thing. Dmitrii Mendeleev and the Shadow of the Periodic Table. Basic Books, New York. Ch. 5.

17. For perspectives on Beilstein's life and work, see: (a) Gordin MD (2003/4). Beilstein unbound: unraveling the Handbuch der organischen Chemie." Chem. Heritage 21(4):10-11, 32-36. (b) Gordin MD (2005). Beilstein unbound: the pedagogical unraveling of a man and his Handbuch. In Kaiser D (ed.) Pedagogy and the practice of science: historical and contemporary perspectives. MIT Press: Cambridge, MA. pp 11-39.

18. Butlerow A (1861). Z Chem 4:549–560. A full English translation of this paper is available: Kluge FF, Larder DF (1971) J Chem Educ 48:289-291.

19. Hargittai I (2011). Aleksandr Mikhailovich Butlerov and chemical structure: Tribute to a scientist and to a 150-year old concept. Struct Chem 22:243-246.

20. Most of the information about the educational system and the academic career path in Russia is taken from: Ipatieff VN (1946). The Life of a Chemist. Memoires of Vladimir N. Ipatieff. Stanford University Press, Stanford, CA. Ipatieff's professional career in Russia lasted from the last two decades of the nineteenth century until his defection to the United States in 1930.

Chapter 3
The Rise of Organic Chemistry in Russia: Kazan' and St. Petersburg

3.1 Kazan' and St. Petersburg

During the early part of the nineteenth century, two major locations were largely responsible for the progress of organic chemistry in Russia: St. Petersburg, which had been a center of higher learning since the founding of the Academy of Sciences in 1725, and Kazan' (Fig. 3.1), a city on the Volga River. The emergence of organic chemistry in St. Petersburg is not particularly surprising, given the resources available in the capital and the existence of the Academy of Sciences there, but the same cannot be said for Kazan', a city some 600 miles east of Moscow on the Volga River. Kazan' is now the capital of Tatarstan, in the Russian Federation, but at the turn of the nineteenth century, it was simply an eastern outpost of European Russia. In fact, when Kazan' was awarded a university in 1804 as a result of the 1803 decree of Tsar Aleksandr I—the same decree that established universities in the Baltic-Russian city of Dorpat (now Tartu, in Estonia), St. Petersburg, and the Ukrainian-Russian city of Khar'kov (now Kharkiv), and that also expanded Moscow University—the response of the local population was tepid, at best.

Kazan' had received one of the first two Gymnasia in Russia during the reign of Empress Elizaveta Petrovna [1]. The Gymnasium was affiliated with Moscow University, so many of the local population saw no need for a university at Kazan' as well. But... a university was decreed, so a university was established. Teaching at the university was divided into four faculties: Physics–Mathematics, Law, Medicine, and History–Philosophy. Little could it have been predicted at the time of its foundation that it would be Kazan' where Russian organic chemistry would first blossom, or that the movement of organic chemists from Kazan' to St. Petersburg would lead to the formation of strong intellectual ties between these two widely disparate locations. In fact, the movement of chemists from Kazan' to the capital resulted in the Kazan' chemists' strong traditions of experimental and theoretical chemistry being passed on to St. Petersburg, and to the emergence of a strong school of chemistry in the capital during the 1850s.

D. E. Lewis, *Early Russian Organic Chemists and Their Legacy*,
SpringerBriefs in History of Chemistry, DOI: 10.1007/978-3-642-28219-5_3,
© The Author(s) 2012

Fig. 3.1 Kazan' city in 1828 from a drawing by K. K. Klaus. The university building is the *white* building on the hill at *right*

3.2 Kazan' and the Frontier Chemists

3.2.1 Introduction: Kazan' University

There is a level of consensus among historians of Russian organic chemistry that the beginnings of Russian organic chemistry are best traced to Kazan'. The city itself is ancient, having celebrated its millennium in 2005. It is generally believed to have been founded as a military outpost by the Volga Bulgars, but its strategic position for trade made it a major trading center along the Volga River. In 1455, it became the capital of a powerful Tartar khanate in 1445 under Bulgar Prince Makhmudek, and remained as such until its conquest by Tsar Ivan IV (Ivan the Terrible) in 1552 after a seven-week siege. In 1557, the population of Kazan' was 7,000; when Kazan' university was founded, the population was 40,000; today, the city is an important economic, educational, and cultural center with a population over a million.

As noted above, in contrast to the other three new university centers, Kazan' lacked an upper middle class that would enthusiastically embrace a university. In addition, it was isolated from western Russia—the first bridge across the Volga River, and the rail link with Moscow were only completed in 1896, making communications with the west difficult. Neither of these situations was conducive to the success of a fledgling institution of higher learning. Kazan' was not the location where one would predict the rise of a fine university, let alone what would become, by the end of the nineteenth century, the pre-eminent school of chemistry in Russia, and one that produced almost half the Professors of Chemistry in the Russian Empire.

As the economic center of the Volga and Kama River basins, Kazan' was effectively the easternmost outpost of European Russia. In the eyes of most Russians at the time, however, Kazan' was not really a European city, but an Asiatic one, which makes the rise to eminence of Kazan' University even more remarkable. In the eyes of "European" Russians, this Asiatic city was a less desirable and less prestigious location than any of the other three cities where new universities were to be located. This is one of the reasons why the faculty roster was filled only slowly after its founding: for over a decade the university was only half-staffed, functioning more as an extension of the Kazan' Gymnasium than as an independent institution of higher learning. The other was the mismanagement of the new university by the University Curator (S. Ya. Rumovskii), who was based in St Petersburg, and the Kazan' Gymnasium Director (I. F. Yakovkin), who supplied him with reports that stated that the university should come under the direction of the Gymnasium, rather than vice versa. The official opening of the university occurred on July 5, 1814— over a decade after its founding [2].

Николай Иванович Лобачевский
(1792-1856) Nikolai Ivanovich Lobachevskii

Despite its less than auspicious beginning, during the middle third of the nineteenth century, Kazan' University boasted some of the most productive, perceptive and creative organic chemists practicing the science, as well as some of the most enlightened administration of the time—the mathematician Nikolai Ivanovich Lobachevskii (Николай Иванович Лобачевский, 1792–1856), the eminent mathematician and the developer of non-Euclidean geometry, served as Rector from 1827 to 1846. During 1834–1837, Lobachevskii supervised the construction of a new science building, with the chemistry floor modeled on the Giessen model [2].

The first professors at Kazan' were Germans, who generally taught in Latin [3], as happened at all the new universities founded by Aleksandr I, but this situation gradually changed over the next four decades [4]. Although the universities flourished at first, their status reversed as Aleksandr the young reformer gave way to Aleksandr the reactionary and tyrant. By the end of the Napoleonic wars in 1815, Aleksandr had completely abandoned his liberal ideas.

He placed the universities under the control of inspectors, whose job it was to root out godlessness and revolutionary philosophy. At Kazan', the inspector was Mikhail Leontevich Magnitskii (Михаил Леонтевич Магнитский, 1778–1844), a fanatic who had served as Governor of Simbirsk, and who served in the Ministry of Ecclesiastical Affairs and Education. Magnitskii quickly characterized the university as a hotbed of godless rationalism, but there was almost certainly a degree of personal animus in his assessment: two years earlier, an official in his administration in Simbirsk had applied for a professorship at Kazan', only to be turned down. It was then that Magnitskii began his campaign against the university. In April, 1819, he proposed the drastic step that the university be closed entirely, and the building ceremoniously razed [5]. This was not the radical position that it might seem because, at that time, Kazan' University had been badly mismanaged, and was far from achieving the status required by its founding charter (although it was not alone in this).

But Aleksandr did not assent to the closure of the university. He did, however, confer absolute control over the university on Magnitskii, and the fanatical inspector used this power as a blunt object. He eliminated geology from the curriculum, and restricted the teaching of physics. Professors of physics were to confine their teaching to what could be observed; interpretations and theories explaining what had caused the phenomena that had been observed were prohibited. Magnitskii saw such interpretation as the type of philosophy that the Tsar viewed as fomenting rebellion.

During Magnitskii's purge of the University, religion (that is, *orthodox* Russian Christianity) was the only way to avoid the inspector's axe: By delivering an address, "The Use and Misuse of Natural Science and the Need to Base It on Christian Devotion," [6] the chemistry instructor, Ivan Ivanovich Dunaev, saved his position and began a dramatic rise within the university. In 1835, this same Dunaev, who is described in modern histories as the "unimpressive" professor of chemistry, was forcibly retired from his position.

3.2.2 Zinin and Klaus: The Founding Fathers of the Kazan' School of Chemistry

Николай Николаевич Зинин (1812-1880)
Nikolai Nikolaevich Zinin

Карл Карлович Клаус (1796-1864)
Karl Karlovich Klaus (Carl Ernst Claus)

The scientific growth of the chemistry school at Kazan' started under Nikolai Nikolaevich Zinin (Николай Николаевич Зинин, 1812–1880) [7] and Karl Karlovich Klaus (Карл Карлович Клаус, 1796–1864) [8], neither of whom was trained in chemistry: Zinin's training was in physics and mathematics, and Klaus' training was in pharmacy and botany. Zinin occupied the Chair (*kafedra*) in Chemical Technlogy, and Klaus occupied the Chair in Chemistry. During their time at Kazan', both men were to make seminal contributions to the development of chemistry—Zinin in organic chemistry, and Klaus in inorganic chemistry (although his work has actually had a profound impact on modern organic chemistry).

3.2.2.1 Nikolai Nikolaevich Zinin and His Chemistry

Nikolai Nikolaevich Zinin was born in Shusha, now in Azerbaijan, but on the death of both parents while he was still young, he was sent to live with his uncle in Saratov. He graduated from the Gymnasium here, and then, because the sudden death of his uncle left him without a fortune, he gave up his plans to enter the prestigious St. Petersburg School of Engineering and Communication, and entered the closest university—Kazan'—instead. He took his *kandidat* degree in physics and mathematics from Kazan' in 1833 under Lobachevskii, with a dissertation on the perturbation of the elliptical motions of planets [9]. He took the examinations for the master's degree in physical–mathematical sciences in 1836, and was immediately appointed as adjunct in physics and mathematics. In 1837, he was appointed to the Chair of hydrostatics and hydrodynamics. However, instruction in chemistry was in limbo following the dismissal of Dunaev, and the needs of the university took precedence over Zinin's preferences or training. The faculty council determined that Zinin should specialize in chemistry, so his career path changed forthwith, beginning with his 1836 dissertation for the degree of M. Nat. Sci. [10]. Following the successful defense of this dissertation, Zinin was appointed to a position as adjunct in chemistry in 1837.

This procedure—appointing someone to teach in a discipline despite their lack of formal training in the subject—was common in Russia at the time: the training and preferences of the faculty member were of secondary importance in the face of the needs of the University for someone to teach a subject. The origin of this policy may lie in the fear of Tsar Nicholas I that foreigners (and foreign professors, in particular) were potential agitators, which led to his attempts to reduce the number of foreign professors at Russian universities, already alluded to in Chap. 2.

Whatever the root cause, Zinin would teach chemistry. There was no reason to expect that the change in Zinin's career would lead to eminence as a chemist, and especially not to international stature, but, as Zinin's later career showed, it succeeded beyond anything that could have reasonably been expected. The same situation occurred two decades later with the appointment of Zinin's student, Aleksandr Mikhailovich Butlerov (1828–1882), to teach chemistry despite his having graduated in entomology.

They had appointed Zinin to teach chemistry without him ever having seriously studied the subject, but the Ministry of Education was not unaware of the limitations that this would impose on his teaching, so the young man was sent on a study abroad (*komandirovka*) to western Europe immediately after graduating into the M. Chem. degree. The concept of the *komandirovka* was a result of the Tsar's failed attempt at establishing a Russian pedagogical academy capable of educating university professors at the high levels of competence necessary for them to teach at the university level. The necessary training was only available in western Europe, so promising young scholars were sent abroad to complete their training. Still, most students taking *komandirovky* during the reign of Nikolai were closely monitored by his secret police.

The purpose of Zinin's *komandirovka* was simple: he was to attend the lectures of eminent western European chemists, take notes of those lectures, and then return to Russia and teach from those notes. It was not intended for Zinin to undertake original research in any branch of chemistry. Nevertheless, Zinin turned it into a very productive research experience in organic chemistry. Young Zinin spent three years abroad (1838–1841) in Germany, France (where he attended the lectures of Wurtz) and Britain (where he met Lyon Playfair). He spent the final year (1840–1841) in the laboratory of Justus von Liebig, then arguably the most eminent organic chemist in all Europe, and here he began his research career. Zinin's experience in Liebig's laboratory profoundly affected him [11], and he resolved to institute laboratory instruction in chemistry along the lines of the Giessen model on his return to Russia. This he did, both at Kazan', and later at St. Petersburg.

On his return to Russia, Zinin stopped in St. Petersburg to sit the examinations for the Dr. Chem. degree; he passed the examinations, and he then returned to Kazan' to write and defend a dissertation describing the work he had carried out in Liebig's laboratory [12]. This gained him the Dr. Chem. degree from St. Petersburg University in 1841.

Although he had been destined for the *kafedra* in Chemistry before his *komandirovka*, during his absence that chair had been filled by Klaus, so the vacant chair at Kazan' was the *kafedra* in Chemical Technology. Consequently, when he successfully defended his Dr. Chem. dissertation in 1841, he was appointed Professor of Chemical Technology at Kazan', although he did not want to occupy a chair in technology. In fact, after his graduation he had requested the opportunity to occupy the vacant chair of Chemistry at Khar'kov University. For the next seven years, he remained at Kazan', mainly because the Ministry of Education needed him there to take advantage of what he had learned during his *komandirovka*. However, Zinin disliked his appointment in technology intensely, since his interests were in pure organic chemistry, not chemical technology, and he was constantly applying for a transfer to St. Petersburg. Although his teaching responsibilities were in chemical technology, Zinin's focus was clearly on organic chemistry. He taught only one hour of Chemical Technology per week, and spent the rest of his time teaching other chemistry classes more in line with his interests [13].

Zinin eventually realized that he would not be granted the transfer he sought while he was employed by the Ministry of Education, so he resigned his position in the

Ministry of Education, and applied for an appointment to the Ministry of War. He obtained this appointment in 1848, and was immediately transferred to St. Petersburg to take the Chair in Chemistry at the Medical-Surgical Academy. While in this position, in 1855, he had the opportunity to work with a young man whom he introduced to nitroglycerin: the same young man—Alfred Nobel—made a fortune by stabilizing the explosive, and bequeathed to the world the five original Nobel Prizes.

Zinin held this position until 1864, and was then appointed to be Director of the Chemical Laboratory, a position he held until 1874. In 1865, Zinin was elected Full Academician in the Russian Academy of Sciences. During 1867–1868, he played an active role in the formation of the Russian Physical–Chemical Society, and served as its President for the first ten years (1868–1877).

When he left for western Europe, Zinin had had neither training nor experience in chemistry. The capstone experience of his *komandirovka* was his year in the laboratory of Justus von Liebig. Zinin entered Liebig's laboratory during the period of time that Liebig and Wöhler were carrying out their seminal investigations of benzoyl compounds [14]. The work that Zinin did on oil of bitter almonds led to the discovery of the catalysis of the benzoin condensation by cyanide ion [15], a process that significantly improved on the first method for preparing this compound from oil of bitter almonds.

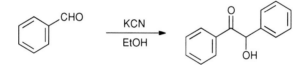

It is not known how Zinin's improvement of this reaction was discovered, but it is not difficult to infer that in an attempt to prepare oil of bitter almonds from benzaldehyde, the inexperienced Zinin may have added the cyanide to the benzaldehyde too slowly, and thus obtained benzoin. During the same *komandirovka*, Zinin carried out the nitric acid oxidation of benzoin to benzil.

The focus of Zinin's work shifted on his return to Kazan', but he resumed his interest in the chemistry of benzoyl compounds after he moved to the Medical-Surgical Academy of St. Petersburg [16]. In fact, he pursued this line of research for the remainder of his career. It was at Kazan', however, that his greatest discovery was made.

The discovery was partly serendipitous because Zinin could not pursue his intended studies on oil of bitter almonds—the importation of this highly toxic material (it is benzaldehyde cyanohydrin) into Russia was prohibited at the time.

He began a study of the reactions of hydrogen sulfide with a variety of organic compounds, including nitrobenzene and the nitronaphthalenes. In these reactions, he obtained products that he called benzidam and naphthylidam [17]. In a footnote at the end of Zinin's paper in the *Journal für praktische Chemie*, Carl Julius Fritzsche (Yulii Fedorovich Fritsshe, Юлий Фёдорович Фрицше, 1808–1871),[1] then working at the Academy of Sciences in St. Petersburg, pointed out the identity of Zinin's oil, benzidam, with the compound he had called aniline in his 1840 paper on the products of treating indigo with caustic base [18]. Within five years, Zinin had also reported the preparation of azoxybenzene, azobenzene, and benzidine (via hydrazobenzene, an easily oxidized compound that he attempted to purify and stabilize by treating it with sulfuric acid to convert it to the much less easily oxidized hydrogen sulfate salt) [19].

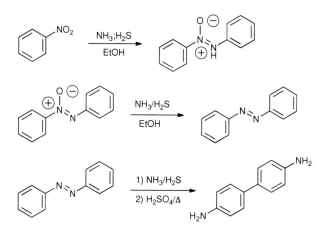

Zinin's reduction of nitrobenzene to aniline had a dramatic impact on the development of aromatic chemistry. Zinin made his discovery at the beginning of the intensive investigation of coal tar by Hofmann, who quickly recognized its importance. During the nineteenth century, textiles dominated the manufacturing

---

[1] Fritzsche was born in Neustadt, and studied under Mitscherlich from 1829–1831. He took his Ph.D. in botany from Berlin in 1833, but his time as an assistant to Mitscherlich steered him to a career in chemistry. He moved to Russia in 1834, where his services were retained by the Russian government. His career was entirely in organic chemistry, with much of his work being concerned with heterocyclic aromatic nitrogen compounds (murexide, uric acid, and indigo), and in the hydrocarbons of coal tar, from which he isolated chrysene, pure antharacene, and retene (7-isopropyl-1-methylphenanthrene). In 1838 he was appointed to the Academy of Sciences as Adjunct, and became a full Academician in 1852; after completion of the laboratory facilities at the Academy in 1866, he shared facilities with Zinin. He always enjoyed excellent health, but in 1869 he became afflicted with a stroke that left one side paralyzed, and robbed him of his once fluent speech and excellent memory. In 1870, he returned to Germany to seek help for his condition, to no avail. For more biographical details, see: (a) Harcourt AV (1872) Anniversary Meeting, March 30th, 1872. J Chem Soc 25:341–364 (esp. 345–348); (b) Sheibley FE (1943) Carl Julius Fritzsche and the Discovery of Anthranilic Acid, 1841. J Chem Educ 20:115–117.

industry, and Perkin's discovery of mauveine in 1856 resulted in a high demand for aniline derivatives [20]. Aniline, which had been prepared earlier by destructive distillation of indigo with base [21], is a key compound in the production of synthetic dyes, and its simple production by reduction of the aromatic nitro compounds revolutionized aromatic chemistry. In his nekrolog of Zinin, August Wilhelm von Hofmann assessed Zinin's contribution as "epoch-making," and described its impact thus: "If Zinin had done nothing more than to convert nitrobenzene into aniline, even then his name should be inscribed in golden letters in the history of chemistry" [22]. Zinin's failure to capitalize on his discovery may have been symptomatic of the organization of chemical and technological education in Russia at the time, as has been discussed by Brooks [13].

3.2.2.2 Karl Karlovich Klaus and His Chemistry

Klaus childhood can only be described as traumatic: lonely and unloved. He was born in Dorpat to a talented painter who died when Klaus was four years old. His mother married again (another artist), but she died when Klaus was five years old. When his stepfather also married again, Klaus, who showed significant talent as an artist, became a neglected child. At fourteen, after failing to complete the course of study at the Gymnasium, Klaus was forced to earn his own living. He became apprenticed to an apothecary in St. Petersburg, and began a process of self-education, reading books on pharmacy, chemistry, and related sciences. He was so successful at this that, despite his lack of formal education, he was able to pass the examinations at the Military Medical Academy of St. Petersburg to become first, assistant pharmacist, and then *provisor*, the qualification required for advanced study in pharmacy (this degree was later re-named as the *kandidat* degree in pharmacy) [23]. In 1815, he returned to Dorpat to pass the examinations in pharmacy at the University, and then returned to the St. Petersburg apothecary. In 1817, he moved east to Saratov, as *provisor* of pharmacy, but the study of the flora and fauna of the Volga steppes was what really consumed him [24]. In 1821, he married, and in 1826 he opened his own pharmacy in Kazan'. He returned to Dorpat in 1828 to begin his formal education. In 1831, he became an assistant in the chemical laboratory, and graduated with the degree of *kandidat* of scientific philosophy in 1835, and in 1837, he defended his dissertation for the *magistr* of philosophy degree in the area of analytical phytochemistry [25]. In 1839, he presented his Dr. Chem. dissertation at Kazan' on the subject of mineral waters of the Sergiev district of the Samara region [26].

In 1837 Klaus presented a trial lecture at the Medical-Surgical Academy in St. Petersburg as part of his application for an academic position. This led to his appointment as an assistant in the *kafedra* of Chemistry at Kazan' and as Director of its chemical laboratory. After graduating into the Dr. Chem. degree in 1839, he was promoted to Extraordinary Professor of Chemistry at Kazan', and, in 1844, he was promoted to Ordinary Professor. He remained at Kazan' until 1852, when he returned to Dorpat as Professor of Pharmacy.

Klaus' lasting fame is due to his isolation and identification of ruthenium, which capped an impressive body of work in the chemistry of the platinum metals [27]. While this means that his major contributions to the development of chemistry in Russia were in inorganic chemistry (he wrote a book on the chemistry of the platinum metals), he and his students also undertook the study of the reactions of organic compounds (especially natural oils) with platinum metal compounds. In 1842, Klaus published the results of the reactions of halogens with camphor in phosphorus halides as solvent [28], and a decade later Aleksandr Mikhailovich Butlerov carried out the first reaction of an olefinic hydrocarbon with osmium tetroxide while a student at Kazan' under Klaus' direction [29].

3.2.3 Establishing a Pre-eminent School of Chemistry at Kazan': Aleksandr Mikhailovich Butlerov

Between them, Klaus and Zinin were responsible for the training of the most influential Russian organic chemist of the next generation, Aleksandr Mikhailovich Butlerov (Александр Михайлович Бутлеров 1828–1886). Under Butlerov, the stature of the Kazan' school of chemistry grew, making it surpass both Moscow and St. Petersburg. Not only was Butlerov an eminent organic chemist in his own right, but three of his students also became eminent Professors of Chemistry during his lifetime, and eight of their students, in turn, occupied Chairs of Chemistry by the turn of the twentieth century; practically all of Butlerov's students and their students have gone into the text-books as the developers of name reactions or rules. Butlerov's influence remains unparalleled in terms of mentoring students who became eminent chemists in their own right.

Aleksandr Mikhailovich Butlerov was born to a retired lieutenant colonel in Chistopol. His family was part of the minor nobility, and owned part of the village of Butlerovka. After graduating from the Gymnasium in Kazan', Butlerov entered Kazan' university, where he was a student from 1844 to 1849. He studied chemistry under Zinin and Klaus, but after Zinin's departure for St. Petersburg in 1848, he turned his focus back to another of his interests: entomology. In 1849, he presented his *kandidat* dissertation on the diurnal butterflies of the Volga region [30]. Despite his *kandidat* degree being in entomology, Zinin's departure meant that the university needed Butlerov to teach chemistry. He began teaching chemistry part-time in 1849, and then, from 1850, as Klaus' full-time assistant. He completed the dissertation for the degree of M. Chem. in 1851, presenting a dissertation on the oxidation of organic compounds that was largely a historical review, with little new chemistry or original ideas [31]. On graduating with his master's degree, Butlerov was appointed Extraordinary Professor of Chemical technology. In 1854, he received his Dr. Chem. degree from Moscow University for another largely historical work [32] (the dissertation had been presented at Kazan' University, but one of the three examiners did not find it of sufficient merit for the award of the degree). It is worthwhile noting that neither dissertation was published. Following Klaus' departure to Dorpat in 1852, the responsibilities for all chemistry teaching devolved upon Butlerov.

In 1857, Butlerov was sent on a *komandirovka*, where he spent time in Germany and France. During the trip, he became acquainted with Kekulé and Erlenmeyer, and he spent six months in the Paris laboratory of Adolphe Wurtz. These were heady years for organic chemistry, with the new structural theory beginning to take shape in Germany [33] and Paris [34], and Butlerov returned from his trip as one of the new theory's most ardent adherents. Less than a decade earlier, he had been a strong adherent of the dualistic theories of Berzelius; he and Klaus taught from Löwig's textbook. However, in 1854, Zinin encouraged him to familiarize himself with the work of Gerhardt and Laurent, and as a result, Butlerov quickly abandoned his obsolete views and became one of their adherents. This was undoubtedly a major factor contributing to his receptivity to the modern structural ideas that were being developed in western Europe during his *komandirovka*.

On his return to Russia in 1858, Butlerov was promoted to Professor. He installed gas in the chemical laboratory at Kazan', and expanded it. He instituted a policy, which he carried with his to St. Petersburg, that all students were required to complete a specified curriculum of practical work. This was an important change, since it raised the level of instruction in organic chemistry in Russia, putting Russian-educated organic chemists on a much more equal footing with their western counterparts. It is no accident that three of his students from Kazan'—Markovnikov, Zaitsev, and Popov—occupied Professorial Chairs of Chemistry at Russian universities during Butlerov's lifetime.

In March, 1860, he became the last Imperial (i.e. appointed) Rector of Kazan' University, a position he resigned in August, 1861, after coming into conflict with the student body. Against his wishes, he became the first elected Rector of the university in October, 1862; again, internal struggles and conflicts led to his resignation from the post in July, 1863. In 1868, he was appointed to a position as Professor of Chemistry at St. Petersburg University (largely due to the efforts of Beilstein). He did delay his departure until 1869 in order to ensure an orderly transition for the Chair of Chemistry at Kazan'. Butlerov remained at St. Petersburg University for the remainder of his career, retiring in 1885. In 1870, he was elected an Adjunct Member of the Russian Academy of Sciences in the Physico-mathematical section (Chemistry), and was promoted to Associate Academician in 1871, and Full Academician in 1874. Due to the deplorable condition of the laboratory at the Academy, however, Butlerov could not move his experimental work there until the repairs had been completed, in 1882.

Butlerov's major contribution to the development of organic chemistry as a whole is certainly his clarification of the structural theory of organic chemistry [35] first proposed by Kekulé [33] and Couper [34]. As indicated above, Butlerov's *komandirovka* was something of an epiphany: he returned to Russia a committed disciple of the new theory. On his return to Russia, he developed his own version of the theory, which he presented to the Speyer conference of the *Gesellschaft Deutscher Naturforscher und Ärzte* in September, 1861 [36], and continued its development for the next several years [37]. He used his version of the theory to predict the existence of isomers of molecules, and he then immediately set about finding evidence to support it. He included it into his lectures, and in 1864, he wrote the first text book based entirely on the new theory. Five years later, the book was translated into German [38].

Butlerov's synthetic work covered a wide range of areas. During his time in the laboratory of Wurtz, he prepared methylene iodide [39], formaldehyde [40], and hexamethylenetetramine [41]. In 1861, he reported the synthesis of "formose," a mixture of saccharic substances including ribose, by the polymerization of form-aldehyde with calcium hydroxide [42]. A mechanism (Fig. 3.2) was proposed for the reaction by Breslow, in 1959 [43]; more recently, reaction has assumed renewed importance as a possible source of prebiotic carbohydrates [44].

As early as the 1860s, Butlerov suggested that a dynamic equilibrium could occur between structural isomers, presaging the modern concept of tautomerism; during his studies of the polymerization of isobutylene and amylene, he noted the presence of isomers of these hydrocarbons in the product mixture, which led him to formally propose the existence of a dynamic equilibrium between constitutional isomers: the concept now known as tautomerism [45].

As part of his improvements in the utility of structural theory, Butlerov used it to make predictions about chemical structure, and he then set about verifying these

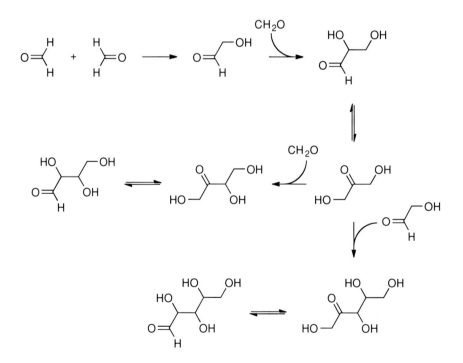

Fig. 3.2 The Breslow mechanism of the formose reaction involves the initial condensation of two molecules of formaldehyde to glycolaldehyde, and subsequent crossed aldol and tautomerization reactions to build the aldose and ketose products

predictions experimentally. This he did in 1863, when he reported the first synthesis of *tert*-butyl alcohol from phosgene and dimethylzinc [46], and, a year later, from acetyl chloride and dimethylzinc [47], albeit in modest yield.

$$H_3C-\overset{\displaystyle O}{\underset{\displaystyle Cl}{C}} \quad \xrightarrow[\text{2) } H_2O]{\text{1) } Me_2Zn} \quad H_3C-\overset{\displaystyle OH}{\underset{\displaystyle H_3C \quad CH_3}{C}} \quad (24\text{-}27\%)$$

Butlerov expanded his alcohol synthesis over the next few years [48] to the synthesis of tertiary alcohols in general from dialkylzinc reagents and carboxylic acid chlorides. Until the discovery of the Grignard reaction in 1900 [49], this represented the state of the art synthesis of tertiary alcohols from carboxylic acid derivatives. As we shall see in subsequent chapters, this reaction was extended further by the next two generations of students in what was known as the "Butlerov School."

3.3 St. Petersburg and the Rise of Organic Chemistry in the Imperial Capital

3.3.1 St. Petersburg University

St Petersburg University, now St. Petersburg State University, was the successor to the Petersburg Pedagogical Institute, founded in 1804, and renamed the Main Pedagogical Institute in 1814. On its renaming in 1814, the Institute, which occupied part of the Twelve Collegia building, consisted of three Faculties: the Faculty of Philosophy and Law, the Faculty of History and Philology and the Faculty of Physics and Mathematics. Since the Pedagogical Institute was itself the successor to the St. Petersburg Academy, it is frequently considered the oldest institution of higher learning in Russia. In 1821, the Institute was renamed the St. Petersburg Imperial University. Unlike Kazan' University, the University did not have trouble filling its faculty ranks, although it is notable that at least two of the most important Professors at St. Petersburg actually had their start at Kazan': Zinin and Butlerov both left their positions at Kazan' to take up appointments at St. Petersburg.

Chemistry at St. Petersburg (Fig. 3.3) was part of the faculty of Physics and Mathematics, with the first lecture and laboratory classes in chemistry being delivered in 1833. With the appointment of Aleksandr Butlerov to the faculty in 1868, the Chemistry Division was split into three: organic chemistry, headed by Butlerov, inorganic chemistry, headed by Dmitrii Ivanovich Mendeleev (Дмитрий Иванович Менделеев, 1834–1907), and analytical chemistry, headed by Nikolai Aleksandrovich Menshutkin (Николай Александрович Меншуткин, 1842–1907). Chemistry did not become a separate, independent department of the university until the Soviet era, in 1929.

Fig. 3.3 The chemical laboratory of the University of St. Petersburg in 1894

3.3.1.1 Aleksandr Abramovich Voskresenskii

Александр Абрамович
Воскресенский (1809-1880)
Aleksandr Abramovich
Voskresenskii

The first Russian chemist of note at St. Petersburg was Aleksandr Abramovich Voskresenskii (Александр Абрамович Воскресенский, 1809–1880), who is known in Russia as the "Grandfather of Russian Chemistry" [50]. Voskresenskii was an organic chemist, but his most important contributions to the development of chemistry in the capital arose from his mentorship of young and upcoming Russian organic chemists. During his tenure at St. Petersburg, Voskresenskii mentored Menshutkin, who became head of the division of analytical chemistry, and who was a pioneer in physical organic chemistry, and most especially, Mendeleev, who became head of the division of inorganic chemistry and who, although he wrote an award-winning textbook of organic chemistry, gained lasting fame as the father of the periodic table of elements.

Voskresenskii, the son of a deacon in the parish church who died when he was only five, was born in Torzhok, in the province of Tver, northwest of Moscow, and educated at the local clerical school "on the Crown's account." His performance was good enough that he was permitted to enrol in the District Clerical Seminary, and he was clearly on the way to an ecclesiastical career. However, this prospect was not appealing to the young man, who enrolled in the Glavnii Pedagogical Institute in St. Petersburg in 1829. Seven years later, he graduated at the head of his class, this while supporting himself through a variety of menial jobs. At Glavnii, Voskresenskii studied under the thermodynamicist, Germaine Henri Hess (German Ivanovich Gess, Герман Иванович Гесс, 1802–1850), and on Hess' recommendation, he was permitted to take a *komandirovka* to Germany. Here he attended lectures by Eilhard Mitscherlich, Heinrich Rose, and Gustav Magnus, but it was at Giessen, where he could study practical laboratory chemistry, that he spent his most productive time. At the time, Giessen had a reputation as a politically "free-thinking" university, and it took Liebig's personal intercession for Voskresenskii to be permitted to stay there instead of being forced to return to Russia. This came at a price: for his entire time at Giessen, Voskresenskii was subject to surveillance by the Tsar's Secret Police. However, it is also likely that Voskresenskii's example during his time at Giessen paved the way for other young Russian chemists to study with Liebig.

Voskresenskii's work at Giessen led to three published papers, one on the effects of sulfuric acid on ethylene [51], one on the composition of naphthalene [52], and his major work on the composition of quinic acid, during which he described the synthesis of benzoquinone [53].

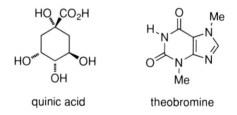

quinic acid   theobromine

Voskresenskii returned to St. Petersburg in 1838, and was appointed Adjunct in the Department of Chemistry at St. Petersburg University. At the same time, he was made Inspector of the Glavnii Pedagogical Institute. He defended his doctoral dissertation in 1839, and was promoted to Extraordinary Professor in 1843, and Ordinary Professor in 1848. In 1861, he was made Dean of the Physical-Mathematics Faculty, and two years later he as elected Rector of the university. In 1864, he was elected a Corresponding Member of the Russian Academy of Sciences, and during the period 1867–1868 he was one of the founding fathers of the Russian Physical–Chemical Society. In 1867, he was appointed Trustee of the Kharkov Teaching District, but this position did not turn out to be compatible with his talents; after numerous battles with those in the Ministry, he resigned in 1869, and retired to his estate near where he was born. Here he remained until his death 11 years later.

After his return to St. Petersburg, Voskresenskii continued his work on quinic acid and began work on the major alkaloid of cacao, theobromine [54]. However, his experimental career was short-lived. Instead, he chose to dedicate his career to teaching chemistry at the highest level. In 1848, he was joined at St. Petersburg by Zinin, and between the two of them, they set a firm foundation for the growth of chemistry at St. Petersburg and in Russia. Both chemists were instrumental in the formation of the Russian Physical–Chemical Society in 1868. As mentioned earlier, two of Voskresenskii's students went on to become leaders in the same division of the university. Based on memoirs written by one of his students, Mendeleev, Voskresenskii was an exacting and inspiring lecturer. His contributions to the development of chemistry in Russia were also not unnoticed by Mendeleev, who wrote, "Voskresenskii and Zinin, his contemporary, share the honor of being the founders of an independent Russian trend in chemistry" [55].

References

1. Seton-Watson H (1967). The Russian Empire, 1801-1917. Oxford University Press, Oxford, p. 35-36.
2. Vinogradov SN (1965). Chemistry at Kazan University in the nineteenth century: A case history of intellectual lineage. Isis 56:168-173.
3. At Kazan' University in 1809, eight subjects were taught in Russian, five in Latin, three in French, and one in German: (a) Ikonnikov VS (1876). Russkie universitety v svyazi s khodom obshchestvennogo obrazovaniya [Russian universities in connection to the progress of public education]. Vestnik Evropy 10:492-550 (p. 549); (b) Vucinich AS (1860). Science in Russian Culture. A History to 1860. Stanford University Press, Stanford, p. 218.
4. An excellent description of the growth of the chemistry department at Kazan', as well as those at Moscow and St. Petersburg, is to be found in Brooks NM (1990). The Formation of a Community of Chemists in Russia, 1700-1870. PhD Diss, Columbia University.
5. Flynn JT (1971). Magnitskii's purge of Kazan University: A case study in the uses of reaction in nineteenth-century Russia. J. Mod. Hist. 43:591-614.
6. (a) Leicester HM (1947). The history of organic chemistry in Russia prior to 1900. J. Chem, Educ. 24:438-443; (b) Menshutkin BN (1921). Nikolai Nikolaevich Zinin, zhizn' i deyatelnosti [Nikolai Nikolaevich Zinin: His life and activities]. Grzhebin Press: Berlin and St. Petersburg.
7. For biographies of Zinin, see: (a) Leicester HM (1940). N. N. Zinin, an early Russian chemist. J. Chem. Educ. 17:303-306; (b) Butlerov AM, Borodin, AP (1881). Nikolaus Nikolajewitsch Zinin. Ber Dtsch Chem Ges 14:2887-2908.
8. For biographies of Klaus, see: (a) Klyuchevich AS (1972). Karl Karlovich Klaus. Kazan'; (b) Weeks ME (1945). Ruthenium. Discovery of the Elements. 5th. edn. Journal of Chemical Education: Easton, PA, pp. 260-268; (c) Menshutkin BN (1928). Karl Karlovich Klaus. Ann. Inst. Platine (Leningrad). 6:1-10; (d) Menschutkin BN (1934). Discovery and early history of platinum in Russia. J Chem Educ 11:226-229.
9. Zinin NN (1833). O perturbatsiyakh ellipticheskogo dvizheniya planet [On perturbations of the elliptical motion of planets. Kand. Diss, Kazan'.
10. Zinin NN (1836). O yavleniyakh khimicheskogo srodtsva i o prevoskhodstve teorii Bertseliusa o postoyannykh proportsiyakh pered khimicheskogo sytatikoi Bertoletta [On the theory of chemical affinity and the superiority of the theory of Berzelius on chemical proportionality constants over the chemical statics of Berthollet]. Diss M Nat Sci, Kazan'.

11. Zinin's reports from his komandirovka: (a) Zinin NN (1837). Report to the Trustee of the Kazan Educational District, 1837. Cited in Polishchuk VR (1981). Chuvstvo veshchestva. Znanie, Moscow, pp. 32-35; (b) Zinin NN (1838). Report to the Trustee of the Kazan Educational District, 1838. Cited in Polishchuk VR (1981). Chuvstvo veshchestva. Znanie, Moscow, pp. 55-57; (c) Zinin NN (1839). Report to the Trustee of the Kazan Educational District, 1839. Cited in Polishchuk VR (1981). Chuvstvo veshchestva. Znanie, Moscow, pp. 57-58.
12. Zinin N (1840). O soedineniakh benzila i ob otrytykh telakh, otnosyashchikhsya k benzoilovomu rodu [Compounds of Benzoyl and the Discovery of New Substances Belonging to the Benzoyl Family]. Dr Chem Diss, St Petersburg.
13. Brooks NM (2002). Nikolai Zinin and synthetic dyes: The road not taken. Bull Hist Chem 27:26-36.
14. Wöhler F, Liebig J (1832). Untersuchungen über das Radikal der Benzoesäure. Ann Chem Pharm 3:249–287.
15. Zinin N (1839). Beiträge zur Kenntniss einiger Verbindungen aus der Benzoylreihe. Ann Chem Pharm 31:329–332.
16. (a) Zinin N (1840)..Ueber einige Zersetzungsprodukte des Bittermandelöls. Ann Chem Pharm 3:186–192; (b) Zinin N (1861). Ueber das Benzil. Justus Liebigs Ann Chem 119:177-179; (c) Zinin N (1861). Ueber das Benzil. J Prakt Chem 82:446-452.
17. Zinin NN (1843). Opisanie nekotorykh novykh organicheskikh osnovanii, poluchennykh pri deistvii serovodoroda na soedineniia uglevodorodov s azotnovatoi kislotoi [Description of several new organic bases obtained through the action of hydrogen sulfide on compounds of hydrocarbons with nitric acid]. Bull Sci Acad Imp Sci St Petersbourg 10:273-285; (b) Zinin N (1842). Organische Salzbasen, aus Nitronaphtalose und Nitrobenzid mittelst Schwefelwassershoff entstehend. Ann Chem Pharm 44:283-287; (c) Zinin N (1842). Beschreibung einiger neuer organischer Basen, dargestellt durch die Einwirkung des Schwefelwasserstoffes auf Verbindungen der Kohlenwasserstoffe mit Untersalpetersäure. J Prakt Chem 27:140-153; (d) Zinin N (1844). Einwirkung von Schwefelammonium auf Nitronaphthalese und Binitrobenzid. Ann Chem Pharm 52:361-362.
18. Fritzsche J (1840). Ueber das Anilin, ein neues Zersetzungsprodukt des Indigo. J Prakt Chem 20:453-459; (b) Fritzsche J (1840). Ueber das Anilin, ein neues Zersetzungsprodukt des Indigo. Justus Liebigs Ann Chem 36:84-90.
19. Zinin N (1845). Ueber das Azobenzin und die Nitrobenzinsäure. J Prakt Chem 36:93-107; (b) Zinin N (1853). Ueber Azobenzid, Azoxybenzid, und Seminaphthadin. Ann Chem Pharm 85:328-329.
20. For a discussion of the rise of the dye and coal-tar industries, see: Aftalion F (1991). A History of the International Chemical Industry. Benfey OT (transl). University of Pennsylvania Press: Philadelphia, ch. 3.
21. (a) Unverdorben O (1826). Ueber das Verhalten der organischen Körper in höheren Temperaturen. Ann Phys 84:397–410; (b) Fritzsche J (1841) Ueber die Produkte der Einwirkung von Kali auf Indigblau. Ann Chem Pharm 39:76-91.
22. Hofmann AW (1880). N. N. Zinin: Nekrolog. Sitzung vom 8 März 1880. Ber Dtsch Chem Ges 13:449-450.
23. For a description of the education process for pharmacists in Russia, see: Möller HJ (1882). Some remarks upon modern pharmaceutical study. Am J Pharm 56:313-323.
24. Claus C (1851). Lokalfloren der Wolgagegenden. Beiträge zur Pflanzenkunde des Russischen Reichs. (Imp Acad Sciences, St Petersburg.
25. Claus C (1837). Grundzüge der Analytischen Phytochemie. M Phil Diss, Dorpat.
26. Klaus KK (1839). Khimicheskoe razozhenie Sergievskikh mineral'nykh vod [Chemical composition of the Sergiev mineral waters]. Dr Chem Diss. Kazan'.
27. (a) Klaus KK (1844). Khimicheskie issledovaniya ostatkov ural'skoi platinovoi rudy i metalla ruteniya [Chemical investigations of the residues of Ural platinum ores and the metal ruthenium]. Sci Writings Kazan' Univ; (b) Claus C (1844). Ueber den Platinrückstand. J Prakt Chem 32:479-492; (c) Claus C (1845). Entdeckung eines neuen Metalls. Ann Phys

140:192-197; (d) Claus C (1845). Untersuchung des Platinrückstandes nebst vorläufiger Ankündigung eines neuen Metalls. Ann Phys 141:200-221.

28. Claus C (1842). Ueber das Verhalten des Camphers zu den Haloïden. J Prakt Chem 25:257-275.

29. Butlerow A (1852). Ueber die Einwirkung der Osmiumsaüre auf organische Substanzen. Ann Chem Pharm 84:278-280.

30. Butlerov AM (1849). Dnevnie babochki Volgo-Uralskoy fauny [Diurnal butterflies of the Volga-Ural region]. Kand Diss, Kazan'.

31. Butlerov AM (1851). Ob okisleny organicheskikh soedineny [On the oxidation of organic compounds]. M Chem Diss, Kazan'.

32. Butlerov AM (1854). Ob efirnykh maslakh [On ethereal oils]. Dr Chem Diss, Moscow.

33. Kekulé A (1858). Ueber die Constitution und die Metamorphosen der chemischen Verbindungen, und über die chemische Natur des Kohlenstoffs. Ann Chem Pharm 106: 129–159.

34. Couper AS (1858). Sur une nouvelle théorie chimique. Comptes rend 46:1157–1160. This paper was translated by Leonard Dobbin as part of a 1953 publication by the Alembic Club, On a New Chemical Theory and Researches on Salicylic Acid. Papers by Archibald Scott Couper; (b) Couper A-S (1858). Sur une nouvelle théorie chimique. Ann Chim [3] 53:469– 489. These papers appeared in English and German: (c) Couper AS (1858). On a new chemical theory. Philos. Mag. 16:104–116; (d) Couper AS (1859). Ueber eine neuer chemische Theorie. Ann Chem Pharm 110:46–51.

35. Monographs: (a) Benfey OT (1964). From Vital Force to Structural Formulas. In Hart H (ed) Classical Researches in Organic Chemistry. Houghton-Mifflin: New York; (b) Lewis DE (2010). 150 Years of Organic Structures. In Giunta CJ (ed) Atoms in Chemistry: From Dalton's Predecessors to Complex Atoms and Beyond. ACS Symp Ser, American Chemical Society: Washington, D.C. 1044:35-57.

36. Butlerow A (1861). Einiges über die chemische Structur der Körper. Z Chem 4:549-560.

37. Boutlerow, A. (1864). Sur les explications différentes de quelques cas d'isomérie. Bull Soc Chim Paris Nouv Sér 1:100-128.

38. (a) Butlerov AM (1864). Vvedenie k polnomu izucheniyu organicheskoy khimy [An Introduction to the Complete Study of Organic Chemistry]. Kazan'; (b) Butlerow A (1869). Lehrbuch der organischen Chemie: zur Einführung in das specielle Studium derselben. Quandt & Händel: Leipzig.

39. Boutlerow A (1861). Recherches sur l'iodure de méthylène. Comptes rend 46:595-597; (b) Butlerow A (1858). Ueber das Jodmethylen Justus Liebigs Ann Chem 107:110-112.

40. Boutlerow A (1859). Sur le dioxyméthylène. Comptes rend 49:137-138.

41. Butlerow A (1859). Ueber einige Derivate des Iodmethylens. Justus Liebigs Ann Chem 111:242-252.

42. Boutlerow A (1861). Formation synthétique d'une substance sucrée. Comptes rend 53:145–147.

43. Breslow R (1959). On the Mechanism of the Formose Reaction. Tetrahedron Lett 22–26.

44. (a) Orgel LE (2000). Self-organizing biochemical cycles. Proc Natl Acad Sci USA 97:12503–12507; (b) Lambert JB, Gurusami-Thangavelu SA, Ma K (2010). The Silicate-Mediated Formose Reaction: Bottom-Up Synthesis of Sugar Silicates. Science 327:984-986.

45. Butlerow A (1876). In Die V. Versammlung russischer Naturforscher und Aerzte in Warschau 31. Aug./12. Sept.–9/21. Sept. 1876. Ber Dtsch Chem Ges 9:1605; (b) Butlerow A (1877). Ueber isodibutylen. Justus Liebigs Ann Chem 189:44-83.

46. Boutlerow A (1863). Sur quelques composés organiques simples. Bull Soc Chim Paris 582-594.

47. Boutlerow A (1864). Sur l'alcool pseudobutylique tertiare ou alcool méthylique triméthylé. Bull Soc chim Paris 7:106-116; (1866). Sur les alcools tertiares. Bull Soc Chim Paris Nouv Sér 5:17-33.

48. (a) Butlerow A (1863). Studien über die einfachsten Verbindungen der organischen Chemie. Z Chem Pharm 6:484-497; (b) Butlerow A (1864). Über den tertiären Pseudobutylalkohol (den trimethylirten Methylalkohol) Z Chem 7:385-402; (c) Butlerow A (1864). Berichtigung

zur Abhandlung, 'Über den tertiären Pseudobutylalkohol.' Z Chem 7:702; (d) Butlerow A (1865). Über die tertiären Alkohole. Z Chem 8:614-618; (e) Butlerow A (1864). Sur l'alcool pseudobuylique tertiare ou alcool méthylique triméthylé. Bull Soc Chim Fr Nouv Sér 2:106-116; (f) Butlerow A (1865). Über die tertiären Pseudobutyl, oder dreifach methylirten Methylalkohol. Chem Zentralbl 36:168-173. This paper cites ref 47 and: Butlerow A (1864). Bull Soc Chim Paris Nouv Sér. II 106:Août 1864. See also the discussion of Butlerov's work in: Jahresber Fortschr Chem 1863, 475; Jahresber Fortschr Chem 1864, 496.

49. Grignard V. (1900). Sur quelques nouvelles combinaisons organometalliques du magnésium et leur application à des synthèses d'alcools et d'hydrocarbures. C R Hebd Séances Acad Sci Ser C 130:1322-1324; (1901). Action des éthers d'acides gras monobasiques sur les combinaisons organomagnésiennes mixtes, ibid 132:336-338; (1901). Sur les combinaisons organomanésiennes mixtes. ibid. 132:558-561; (1902) Action des combinaisons organomagnésiennes sur les éthers α-cétoniques. ibid 134:849-851; (b) Grignard V (1901). Über gemischte Organomagnesiumverbindungen und ihre Anwendung zu Synthesen von Säuren, Alkoholen und Kohlenwasserstoffen. Chem Zentralbl pt. II:622-625; (c) Grignard V (1901). Sur les combinaisons organomagnésiennes mixtes et leur application à des synthèses d'acides, d'alcohols et d'hydrocarbures. Ann Chim [vii] 24:433-490; (d) Tissier L, Grignard V (1901). Sur les composés organométalliques du magnésium. C R Hebd Séances Acad Sci Ser C 132:835-837; (1901) Action des chlorures d'acides et des anhydrides d'acides sur les composés organo-metalliques du magnésium. ibid 132:683-685; (e) Grignard V, Tissier L (1902). Action des combinaisons organomanésiennes mixtes sur le trioxyméthylène: synthèses d'alcools primaires, C R Hebd Séances Acad Sci Ser C 134:107-108; errata, ibid., 1260.

50. Steinberg C (1965). Aleksandr Abramovich Voskresenskiĭ. Grandfather of Russian chemists. J Chem Educ 42:675-677.

51. Woskresensky A (1838). Ueber die Einwirkung der wasserfrei Schwefelsäure aus das ölbildende Gas. Ann Pharm 25:113-115.

52. Woskresensky A (1838). Ueber die Zusammensetzung des Naphtalins. Ann Pharm 26:66-69.

53. Woskresensky A (1838). Ueber die Zusammensetzung der Chinasäure. Ann Pharm 27:257-270.

54. Woskresensky A (1842). Über das Theobromin. Ann Chem Pharm 41:125-127.

55. Mendeleev DI (1949). A. A. Voskresenskii. Sochineniya, vol. 15:623-624.

Chapter 4
Russian Organic Chemistry Matures: Emergence of a Russian-Trained Professoriate in Organic Chemistry

4.1 Introduction

The rise of the schools of organic chemistry at Kazan' and St. Petersburg proceeded apace during what may be considered the zenith of the science in Russia, the period between 1855 and 1890. The early part of this period is associated with Butlerov at Kazan', and with Zinin and Borodin at St. Petersburg. However, in the ensuing decades, the students of these chemists made important contributions that made organic chemistry in Russia the equal of, or superior to that being carried out in the western world. Leicester has made the suggestion that were it not for the conservative (non-native Russian) membership of the Academy of Sciences, the rise of chemistry in Russia would have continued even longer [1]. The elevation of the standard of Russian chemistry during this period was accompanied by the rise of regional universities, but, more especially, by the rise of chemistry at Moscow University, to join the productive programs in St. Petersburg and Kazan', during the last quarter of the century.

4.2 St. Petersburg

4.2.1 The Imperial Medical-Surgical Academy

In addition to St. Petersburg University and the Imperial Academy, organic chemistry also began to flourish in St. Petersburg during the early part of the nineteenth century at the Petersburg Medical-Surgical Academy. This institution had been founded Tsar Paul in 1798 as the Army Medical Academy, a place for the education of army doctors. It was renamed the Imperial Medical-Surgical Academy in 1808. Although predominantly for the education of physicians, this institution

D. E. Lewis, *Early Russian Organic Chemists and Their Legacy*,
SpringerBriefs in History of Chemistry, DOI: 10.1007/978-3-642-28219-5_4,
© The Author(s) 2012

played an important role in the development of natural and physical sciences in Russia.

In 1848, Nikolai Zinin had left Kazan', and had taken up the Chair of Chemistry at the Medical-Surgical Academy. He spent the next quarter century of his career there, retiring from his position in 1874. As he had started at Kazan, so he continued at St. Petersburg: During his time at the Medical-Surgical Academy, Zinin made it his job to raise the level of the chemistry laboratories, and he was instrumental in planning a new Institute of Chemistry. Under Zinin, the Medical-Surgical Academy became so well known for chemistry that local wags occasionally suggested that it be renamed the Medical-*Chemical* Academy.

4.2.2   The Next Generation of Organic Chemists at St. Petersburg

The rise of chemistry as a significant discipline at St. Petersburg is reasonably attributed to Zinin and Voskresenskii, who educated three of the four individuals to whom may be attributed the consolidation of the position of the science of chemistry in St. Petersburg: Nikolai Aleksandrovich Menshutkin and his friend, Dmitrii Ivanovich Mendeleev, at St. Petersburg University, and Aleksandr Porfir'evich Borodin (Александр Порфирьевич Бородин, 1834–1887), at the Medical-Surgical Academy. The fourth was Friedrich Konrad Beilstein (Fyodor Fyodorovich Beil'shtein, Фёдор Фёдорович Бейльштейн, 1838–1906), who followed Mendeleev into the Chair of Chemistry at the Imperial Technical Institute in St. Petersburg. Both Menshutkin and Mendeleev had been students of Voskresenskii at the St. Petersburg Pedagogical Institute, but Beilstein's path to a leadership position in Russian organic chemistry was significantly different. Even so, the careers of Beilstein and Mendeleev are inextricably linked, so we will treat them together. But we will begin with Menshutkin.

4.2.3   Nikolai Aleksandrovich Menshutkin

One of Voskresenskii's earliest students, Nikolai Aleksandrovich Menshutkin [2], became Professor of analytical chemistry at St. Petersburg, and was a pioneer in physical organic chemistry. Menshutkin was born in St. Petersburg to a fairly wealthy trader's family. At six years of age, he was sent to the best boarding schools in St. Petersburg, and at age 10, he was enrolled also at the St. Peter German School; he graduated in first place from this school in December 1857, shortly after his 15th birthday. Because he was underage for admission, he was required to pass an examination before he was permitted to enter St. Petersburg University. As a student in the natural science department of the Physico-mathematical faculty of the university, Menshutkin received his education in inorganic and analytical chemistry from Voskresenskii, and his education in organic chemistry from Nikolai Nikolaevich Sokolov (Николай Николаевич Соколов, 1826–1877),[1] for whom Menshutkin had great respect and affection [3].

Николай Александрович
Меншуткин (1842-1907)
Nikolai Aleksandrovich
Menshutkin

In 1862, Menshutkin successfully defended his dissertation for the degree of *kandidat* and then immediately took a *komandirovka* to western Europe. He began at the Tübingen laboratory of Adolph Strecker, then moved to the Paris laboratory of Charles Adolphe Wurtz in 1864, and finally to the Marburg laboratory of Hermann Kolbe in 1865. He returned to Russia in 1866, and immediately wrote up the results of his work on phosphorous acid, receiving the degree of M. Chem. the same year [4]. Three years later, he received his Dr. Chem. degree for work on ureides [5]. After obtaining his Dr. Chem. degree, Menshutkin was appointed as Extraordinary Professor of Chemistry at St. Petersburg University and ordinary professor in 1876. In 1879, he was elected Dean of the Physico-mathematical faculty, and he served in this position until 1887; in 1885 he was appointed to the Chair of Organic Chemistry, a position he held until 1902. In 1893, he was given the title of supernumerary professor. Until 1885, Menshutkin had taught analytical chemistry and special courses in organic chemistry, but on assuming the Chair of organic chemistry, he abandoned the teaching of analytical chemistry and devoted himself solely to his specialty. In 1902, he was appointed Professor of Chemistry at St. Petersburg Polytechnic Institute, a position he held until his death. During his time at each institution, Menshutkin supervised the construction and equipping of

---

[1] Sokolov's carer at Odessa will be discussed at more length in Chap. 5.

| | Anfangegeschwindigkeit | | |
	absolut :	*relativ :*	
1. Dimethylcarbinol	26,53	43,85	
2. Aethylmethylcarbinol	22,59	38,10	a
3. Hexylmethylcarbinol	21,19	34,16	
4. Isopropylmethylcarbinol	18,95	31,95	
5. Diäthylcarbinol	16,93	28,86	b

Fig. 4.1 Initial rates of esterification of secondary alcohols as reported by Menshutkin

the chemical laboratories: at the University of St. Petersburg from 1890 to 1894, and at the St. Petersburg Polytechnic Institute from 1901 to 1902. Like his mentors, Menshutkin was one of the founders of the Russian Physical Chemical Society; he served as the editor of its Journal from 1869 to 1900, and as president of the society in 1906. Menshutkin died of a stroke in St. Petersburg just three days after his friend, Mendeleev.

4.2.3.1 The Menshutkin Reaction: Early Physical Organic Chemistry

Menshutkin's lasting contributions to organic chemistry were in the area of physical organic chemistry, where he was a true pioneer. Among his early works were studies of the pyrolyis of amyl acetate, in which he showed the autocatalytic effects of acetic acid, one of the reaction products [6]. Menshutkin also published a long series of papers describing the effects of alcohol structure on the ease of esterification [7]. From these papers, a general appreciation of the effects of structure on the rates of closely-related chemical reactions arose; Fig. 4.1 shows a table of data describing the absolute and relative initial rates of esterification of a series of saturated secondary alcohols.

Menshutkin's name has been preserved in the form of the *Menshutkin reaction*, which now refers to the quaternization of amines with alkyl halides. Using this reaction, Menshutkin definitively demonstrated the potentially dramatic effect of solvent on reaction rates [8], while, at the same time, he extended his studies of the structural effects of the reactants on the rates of chemical reactions [9]. For his pioneering work in physical organic chemistry, Menshutkin received the 1904 Lomonosov Prize—the highest award of the Russian Academy of Sciences.

$$
\begin{array}{c}
R \\
R{-}N \\
R'
\end{array}
\quad \xrightarrow{\ \ R'{-}X\ \ } \quad
\begin{array}{c}
R \ \oplus \\
R{-}N{-}R' \\
R'
\end{array}
\quad X^{\ominus}
$$

Coming out of his kinetic work was Menshutkin's conviction that no reaction can be studied without considering the influence of the solvent [10], a remarkably

modern perspective that was reinforced three decades later by the work of Hughes
and Ingold [11] in their studies of reaction mechanisms.

Александр Порфирьевич
Бородин (1834-1887)
Aleksandr Porfir'evich
Borodin

4.2.4 The Versatile Borodin

Zinin's successor to the Chair of Chemistry at the Medical-Surgical Academy was
his student, Aleksandr Porfir'evich Borodin. During his brief life, Borodin
achieved eminence as both an organic chemist and as a composer [12], although it
has been suggested that the assessment of his accomplishments as a chemist may
have been inflated [13]; the most balanced discussion of Borodin's chemical
accomplishments may be that of Rae [14]. Borodin was the illegitimate son of a
62-year-old Imeretian (Georgian) prince, Luka Stepanovich Gedianov (more
correctly, Gedevanishvili) (1772–1843), and a Russian mother, the 25-year-old
Evdokia Konstantinovna Antonova; he was legitimized by being registered as the
son of Porfiry Ionovich Borodin, his father's valet, making Borodin both his
father's biological son and his father's serf until that same father freed him at
age 7. His mother stayed close to him throughout his life, although she never
recognized him as her son; he referred to her as his "aunt."

Borodin received a fine home education, and he had mastered the French,
German and English languages by an early age. While growing up, he also
exhibited a strong interest in the sciences—botany, zoology, and especially
chemistry—and he also discovered music early: he learned to play the piano, cello
and flute, and by the age of 10 years he had already composed his first work, the
Polka in D minor.

In 1850, he sat for the admissions examinations for the Medical-Surgical Academy and, despite his relative youth (he was barely 16 years old), he was accepted as one of the entering students. At the Academy he studied chemistry under Zinin, with whom he continued to the completion of his M.D. degree in 1858. He became Zinin's favorite student, and was Zinin's choice to succeed him, but Zinin was disturbed by the time he spent on his music, scolding him with, "Mr. Borodin, it would be better if you gave less thought to writing songs. I have placed all my hopes in you, and want you to be my successor one day. You waste too much time thinking about music. A man cannot serve two masters" [15]. Although he was a qualified physician, Borodin became ill at the sight of blood, so he never practiced. As an aside, Borodin's M.D. dissertation had the distinction of being the first at St. Petersburg written and defended in Russian rather than Latin.

Following his graduation, he traveled to Heidelberg on the advice of his doctoral mentor, entering the laboratory of Robert Wilhelm Bunsen. Within a year, however, he transferred to the laboratory of Emil Erlenmeyer (the elder). In 1860, he attended the Karlsruhe conference as a Russian delegate with Mendeleev; at this conference, the atomic weights of the elements were finally fixed, which was absolutely critical to Mendeleev's successful formulation of the periodic table. Later that year, Borodin traveled with Mendeleev and Zinin to southern Europe. From 1860 to 1861, he was in Italy, where he worked in the laboratory of Sebastiano de Luca and Paolo Tassinari in Pisa.[2] While in Italy, he met Ekaterina Sergeevna Protopopova, whom he married in St. Petersburg two years later. Ekaterina, whose health was never robust, was herself a musician, and was said to possess "perfect pitch."

On his return to Russia in 1862, Borodin was appointed docent at the Medical-Surgical Academy (meaning that he received no salary, but received a portion of the fees paid by students to attend his lectures); in 1864 he was appointed Professor of Chemistry at the Academy. In 1872, he helped found courses for women at the Academy, and for much of his professional life he was a strong proponent of the education of women (as an aside, for all Russia's reputation for being backward, the chemistry departments in Russia actually boasted some of the most forward-thinking individuals of the nineteenth century when it came to the education of women). When the medical education of women was halted a decade later under the regency of Aleksandr III, the blow devastated him.

As a professor of chemistry at the Medical-Surgical Academy, Borodin's teaching load was heavy, even by contemporary standards. Consequently, he found that the demands of his teaching left little time for either chemical research or for

---

[2] Paolo Tassinari (1829–1909) was an Italian analytical chemist who took the Chair at Pisa in 1862 after a brief appointment to the University of Bologna, where he taught Analytical, Mineralogical, and Metallurgical Chemistry. Sebastiano de Luca (1820–1880) was a student of Piria, in Naples, and a close friend of Stanislao Cannizzarro. Like him, de Luca was of a revolutionary bent. He served with the rebels in the 1848 revolution, and when their cause failed, he was sentenced to 19 years imprisonment. He escaped apprehension, and fled to France, where he studied with Berthelot. He returned to Italy in 1857, replacing Piria as Chair at Pisa.

composition. He wrote that his friends—both in chemistry and music—often wished him poor health, since it was only when he was too ill to teach that he could accomplish anything in the research laboratory or his composing.

Borodin's musical legacy is also significant, and there are numerous biographies of Borodin, the musician. He remains the only chemist to have won a Tony award: his music provided the score for the musical *Kismet*, which resulted in him winning the Tony award for the best composer of 1954—67 years after his death! Borodin died suddenly of a cardiac anuerysm at the young age of 53 years while attending a fancy dress ball organized by the professors of the Medical-Surgical Academy.

4.2.4.1 Borodin's Chemistry

The brevity of Borodin's career means that his chemical accomplishments were few; they were, however, important. In 1860 his studies were concerned with the chemistry of benzidine, and he has been credited with the first studies of he benzidine rearrangement (although Shine has pointed out that Borodin's contribution was not, in fact, in this area [16]). In 1861 he published an account of the reaction of silver salts of organic acids with molecular bromine [17].

The reaction results in the oxidation of the carboxylate to the acyl hypobromite, which then undergoes homolysis and loss of carbon dioxide to give the alkyl bromide with one carbon atom less than the starting acid. In what must hold the record for a patent examiner missing prior art, Heinz and Cläre Hunsdiecker were awarded a U.S. patent [18] in 1939—78 years after its original publication—for the same reaction, and the reaction entered the textbooks as the Hunsdiecker reaction [19]. It is only since the 1990 s that Borodin's name has been attached to the reaction he first discovered. In 1862 he prepared the first organic fluorine compound by the treatment of benzoyl chloride with potassium hydrogen fluoride [20].

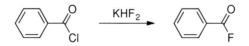

In 1864, he began a project that lasted a decade, and led to the development of what the French chemist, Charles–Adolphe Wurtz (viewed by most—but not all [13]—historians of chemistry as Borodin's competitor) called the aldol reaction [21]. Despite its longer-term association with the name of Wurtz, Borodin's precedence for discovery of this reaction is unambiguous: his first paper describing the reaction of aldehydes with sodium as the base appeared in two papers published in 1864 [22], pre-dating Wurtz' first paper by nearly a decade. Kekulé's first foray into this area appeared in 1869 [23]. Facing competition from these two high-powered competitors, Borodin eventually gave up this line of research.

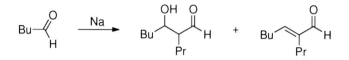

4.2.5 Beilstein and Mendeleev

By far the most famous Russian chemist of the nineteenth century was Dmitrii Ivanovich Mendeleev, the discoverer of the periodic law. Mendeleev did relatively little, however, in the field of organic chemistry. An excellent account of his life and seminal work is contained in the book by historian, Mark Gordin [24], to which the reader is referred. Beilstein, on the other hand, may be the best-known organic chemist that most organic chemists know nothing about, to quote Gordin yet again [25].

Beilstein's relationship with Mendeleev was never good, and it became much worse when, in 1880, Mendeleev was denied the Chair in Technology at the Imperial Academy of Sciences, missing the majority by a single vote (although an extraordinary majority was actually needed for election), only to have Beilstein elected to the same Chair in 1882. And yet, both achieved a high level of scientific eminence: Mendeleev just missed sharing the Nobel Prize for his work on the periodic law, and *Beilsteins Handbuch der Organischen Chemie* continues to be an important reference work for organic chemists over a hundred years after its first appearance.

4.2.5.1 Dmitrii Ivanovich Mendeleev

Like his contemporary, Menshutkin (who actually presented Mendeleev's landmark paper on periodic law because Mendeleev himself was ill), Mendeleev learned his chemistry under Voskresenskii at St. Petersburg. In 1856, he graduated from the St. Petersburg Pedagogical Institute with the degree of *kandidat*, and the same year he presented his dissertation for the degree of M. Chem. This allowed him to take up a position as Docent at St. Petersburg University, but his health forced him to move to southern Russia soon thereafter. Mendeleev's major contribution to organic chemistry was his textbook, *Organicheskaya Khimiya* [*Organic Chemistry*], which he wrote in 1861, and which was based on Gerhardt's view of organic chemistry. Despite the rise of the structural theory of organic chemistry, Mendeleev held to the views of Gerhardt throughout his life. In his book, he focused on similarities in the properties of closely related compounds, but took no stand on the newly-emerged theory of chemical structure; he never accepted the existence of atoms. In 1862 he won the Demidov Prize for this book.

4.2.5.2 Friedrich Konrad (Fyodor Fyodorovich) Beilstein

Beilstein was born in St. Petersburg to an ethnic German family, and although he spoke Russian, he received all his education in German, first at the St. Petersburg German School, then in Germany itself. Following his education at St. Petersburg, in 1853 Beilstein was sent, at the age of fifteen, to Heidelberg, where he studied two years with Bunsen. In 1855, he transferred to Berlin, where he heard lectures by Liebig, and worked under Jolly. Here he completed his first published work, on the diffusion of liquids [26]. In 1856 he returned to Heidelberg, where he met and befriended Hübner and Kekulé, with whom he retained a life-long close friendship. A year after his return to Heidelberg, he moved to Göttingen where, a year later— two days before his twentieth birthday—Beilstein received the Ph.D. for the determination of the structure of murexide [27].

Following his graduation, Beilstein spent a year in the Paris laboratory of Adolphe Wurtz, during which time he studied the action of phosphorus penta-chloride on aldehydes [28]. This work began a continuing thread in his research career involving chlorination reactions of organic compounds. In 1859, he returned to Germany to become Löwig's assistant at Breslau, but Löwig's rigid organization of his research group did not sit well with Beilstein, so when Wöhler offered him a position at Göttingen, he jumped at the chance. In 1860, he returned to Göttingen, and here he continued to carry out research into halogenation of organic compounds.

At the urging of Wöhler and others, who were loath to see one of Germany's brightest young stars lost to Russia, the 1865 offer to Beilstein from St. Petersburg University in 1865, was rapidly countered by the Germans. This meant that it was a huge surprise when, just a year later, Beilstein left Germany for Russia to take up a lower-salary appointment as Professor at the much less prestigious St. Petersburg Technological Institute. Here he remained for the rest of his career. In large part, Beilstein's departure for Russia was prompted by the sudden death of his father, and the needs of his family. That Beilstein viewed the move as permanent is suggested by the fact that, just a year after his return to St. Petersburg, he took the unusual step of giving up his German citizenship and becoming a naturalized Russian subject.

As Mendeleev's successor at the Technological Institute, Beilstein was faced with inadequate laboratories, and with apathetic students who were destined to be engineers, and not scientists. Not only were these students not generally interested in the finer points of theory in chemistry, but neither were the assistants whom Beilstein had to work with. Oddly enough, he might have been well served by taking a leaf out of Löwig's book when it came to running his laboratory at the Technological Institute. As it was, his chemical research output dwindled to practically nothing as the burden of his teaching duties, and his remediation of the laboratory consumed all his time.

In 1880, the Academy failed to elect Mendeleev to the Chair of Technology [29]. The vote of the committee was 12-11 against, but since a two-thirds majority was required for election, the vote was actually not close, and the outcome was recorded as "Conclusion: *not considered elected.*" The ballot that denied Mendeleev even a simple majority was the second vote cast by the chair of the committee (Litke). The decision raised an uproar among Russian scientists and journalists at the time: both placed the blame for Mendeleev's rejection on the "German" party in the Academy. The Secretary of the Russian Physical–Chemical Society, Mendeleev's friend, Menshutkin, proposed that a letter decrying the Academy's action be sent to the newspapers. All the major organic chemists in St. Petersburg signed the letter—except one.

Beilstein's refusal to sign the letter of protest was because he felt that this action was inappropriate, and that the correct forum for dissent was in the form of an address at the next meeting of the Physical–Chemical Society lauding Mendeleev, and criticizing the Academy's actions. This may have, indeed, been the more prudent and more intellectual course of action, but his position was viewed as disloyalty by the "Russian" faction, and from this time on, Beilstein was placed squarely, if unjustly, in the "German" faction. It led to the destruction of his friendship with Butlerov, whose move to St. Petersburg Beilstein had instigated; Beilstein had been instrumental in seeing that Butlerov was appointed to his professorship in the imperial capital. Two years later, the relationship between the two completed its descent into total rancor, when Beilstein himself was elected to the same Chair of Technology that had been denied to Mendeleev. The committee may have elected Beilstein to the Chair by the required two-thirds majority, but Butlerov used the required two-thirds vote at the general assembly of the Academy to block his confirmation: Beilstein's appointment to the Chair of technology was not confirmed until after Butlerov's death.

Beilstein died in October, 2006, becoming the first of three chemical giants of Russia to pass in the space of three months: he was followed in January 1907 by his nemesis, Mendeleev, and then by Menshutkin. Unlike the latter two, however, Beilstein's passing was given scant attention in Russian circles.

4.2.6 Beilstein's Legacy

Beilstein's original research contributions were relatively scant. His studies in Wurtz' laboratory had shown that aldehydes react with phosphorus pentahalides to give alkylidene halides. In 1866, he reported that the chlorination of benzyl chloride gave different results, depending of the temperature of the reaction: at high temperatures, side chain halogenation dominated, while at lower temperatures, nuclear substitution occurred [30]. It was not until the discovery of free radicals by Gomberg [31], over three decades later, and the emergence of free radical reactions in papers by Kharasch [32], and Hey and Waters [33] that the likely reason for the change—a change of mechanism from ionic to free radical—could be proposed. As part of his studies on halogenation, Beilstein devised the elegantly simple Beilstein flame test for halogens using copper wire [34].

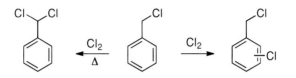

Beilstein's greatest and most lasting contribution on the discipline, however, was his *Handbuch*. He began the *Handbuch* as a textbook in organic chemistry, but it quickly developed into an encyclopedia of organic compounds and their properties. The work involved was monumental, especially since Beilstein insisted that every literature reference be checked before inclusion in the *Handbuch*. He literally read every reference in the first edition, himself. When it appeared, the *Handbuch* appeared in German because of the small market for a work in Russian: a German-language edition would sell many more copies and be read far more widely than a Russian-language edition. Beilstein's decision, however, was seen by his fellow Russian chemists as further evidence of his "German-ness." The judgment was grossly unfair: as editor of the *Zeitschrift für Chemie*, Beilstein had gone to great lengths to see that Russian chemistry was presented to the world in the best possible light—even the text-book of his nemesis, Mendeleev. But Russia had spurned him, so Beilstein naturally looked for a more collegial partner for the continuation of his life's work: he turned to the *Deutsche Chemische Gesellschaft* instead of the *Russkiiskoe Fizicheskoe-khimicheskoe Obshchestvo* [Russian Physical–Chemical Society] when it came time to choose a professional body to continue his work after he was gone.

4.3 The Rise of Organic Chemistry at Moscow

By the middle of the nineteenth century, Russian organic chemistry was poised to enter into an explosive growth phase. Zinin, Klaus and Butlerov had founded a vibrant school of chemistry at Kazan', and Voskresenskii's students at St. Petersburg, Menshutkin and Mendeleev, along with Beilstein at the Technological Institute, had established the foundations of a strong school of chemistry in the imperial capital. The third major location for organic chemistry in Russia—Moscow—was the next to begin the development of a Russian school of organic chemistry.

Moscow University was founded in 1755 by decree of the Empress Elizaveta Petrovna, following the suggestion of Academician Mikhail Lomonosov to Count Shuvalov—a court favorite and the Empress' lover—that a university should be founded in the city. As had been the case at the other major universities in Russia, chemistry at Moscow was initially taught by foreign professors—the inaugural Professor of Chemistry was Johann Kerstens, who had obtained his M.D. at Halle in 1749. Kerstens remained at Moscow until 1770, when he left Russia. Under the foreign Professors of Chemistry, chemistry at Moscow failed to advance as it had at Kazan' and St. Petersburg, but that situation changed in the second half of the nineteenth century. In 1873, a graduate of Kazan' University, Vladimir Vasil'evich Markovnikov (Владимир Васильевич Марковников, 1838–1904) [35], was appointed to the Chair of Chemistry at Moscow University.

4.3.1 Vladimir Vasil'evich Markovnikov

Markovnikov was one of the most colorful and eminent Russian organic chemists of the nineteenth century. He was born to a lieutenant of Chasseurs in Chernorech, a village near Nizhni-Novgorod, and raised in the village of Knyaginino, where his father had inherited an estate. He entered the Gymnasium at age 10, and eight years later, in 1856, he entered Kazan' University as a student in the Financial Division of the Judicial Faculty.

At the time that Markovnikov was a student in economic science at Kazan', students in this course of study were required to take two years of chemistry, as part of a cameral system of education.[3] His first inclination was to study technology, which was taught at the time by Modest Yakovlevich Kittary (Модест Яковлевич Киттары, 1825–1880).[4] Kittary's departure for Moscow in 1859 led to

---

[3] Cameralism was an economic theory prevalent in eighteenth-century Germany. It basically advocated a strong public administration to oversee a centralized, industrial economy. The goals of cameralism were to maximize the efficiency in the ways the state could acquire wealth, and also with the best ways to use that wealth. As Russia moved into the nineteenth century, its industrialization (although slow) led to the belief that the ideas of cameralism would provide the framework to allow the government to ensure that Russia would have a sufficient number of technologically literate bureaucrats to move the nation forward.

[4] Modest Yakovlevich Kittary graduated with the degree of Doctor of Natural Science from Kazan' university in 1844, and in 1853 he was appointed to the Chair of Technology at Kazan' University, where he founded the Kazan' Economic Society and edited its first newsletter. He quickly accumulated a cadre of young technologists, and set about improving the practices in local industry. One of the major Kazan' industries to benefit from Kittary's influence was the Krestovnikov Brothers' plant, which made soap and glycerin pure enough for export. In 1857, Kittary moved to Moscow University as Chair of the Department of Technology that had been established at the urging of local merchants. He remained here as an active educator until his retirement from the university in 1879. Through his writing on aspects of industrial chemistry and technology, Kittary had a major influence on the development of Russian industry during the nineteenth century.

the young Markovnikov coming in contact with Aleksandr Mikhailovich Butlerov, who taught the organic chemistry lecture and laboratory course given in the third year of the cameral science course of study. At the time, Butlerov had just returned from his *komandirovka* in western Europe, and he was actively developing his ideas on structural theory. So Markovnikov had the good fortune to come into contact with Butlerov at the time that Butlerov was developing some of the most advanced theories of the time. Markovnikov was awarded the degree of *kandidat* in the Juridicial Faculty at Kazan in 1860 for his dissertation [36], entitled, "Aldehydes and their relation to alcohols and ketones." He was immediately appointed as an assistant in Butlerov's laboratory, where his research career began. In 1862, he was entrusted with teaching the course in analytical chemistry.

Владимир Васильевич
Марковников (1838-1904)
Vladimir Vasil'evich Markovnikov

In three years, Markovnikov had passed the required examinations for the M. Chem degree, and in 1865 he defended his dissertation [37], "On isomerism in organic compounds," which led to him being awarded a two-year *komandirovka* in western Europe. The title pages of Markovnikov's M. Chem. and Dr. Chem. dissertations are given in Fig. 4.2.

During his time abroad, he attended lectures by Erlenmeyer, Kopp and Kirchoff, and spent brief periods in the laboratories of Erlenmeyer and Baeyer. The bulk of his time he spent with Kolbe at Leipzig, where he followed two earlier Butlerov students (the brothers Zaitsev) into the laboratory.

Markovnikov's situation was different from those of the other students in Kolbe's laboratory because he had entered with an advanced degree in chemistry already, so he was allowed considerable freedom in choosing his topics for research. As he wrote to his mentor, Butlerov [38]:

> My position in Kolbe's laboratory was a little different from all the others, since I was a master *[M. Chem.—DEL]* and had worked on my own subject matter for three years.

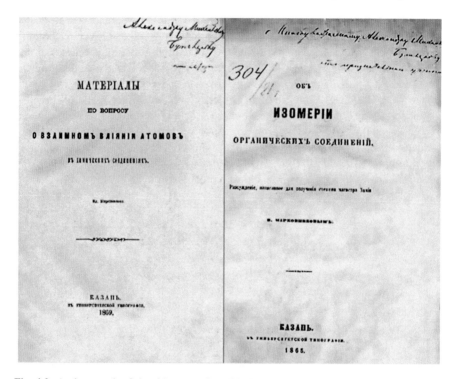

Fig. 4.2 A photograph of the title pages from Markovnikov's Dr. Chem (*left*) and M. Chem. (*right*) dissertations

> During this first year since my arrival in Germany, I have ascertained that, in theoretical matters, the Kazan laboratory has far outstripped all the laboratories of Germany.

Markovnikov studied analytical chemistry in Kolbe's laboratory, but his research work was involved much more with the effects of structure on reactivity. At first glance, it would seem rather ironic that one of the most ardent disciples of the new structural theory of organic chemistry should work with one of its most staunch opponents, and that while there, should develop even more revolutionary ideas based on the new theory. However, it was one of the characteristics of Kolbe—for which he has received relatively scant credit—that he allowed his students to pursue and publish their own lines of research, even when they ran counter to his own opinions (although he occasionally appended long footnotes to those papers outlining his disagreement with their conclusions [39]). While at Leipzig, Markovnikov began to consider (becoming the first to do so) the possibility that atoms in close proximity to each other could influence each other's reactivity. Part of this research was sparked by a question posed to him by Graebe during his time in Baeyer's laboratory: "Why is the chlorine in acetyl chloride different from that in ethyl chloride?" In 1867, Shorlemmer had suggested that the chorination of paraffins would lead to *n*-alkyl chlorides, but Markovnikov demurred, and suggested that secondary alkyl chlorides should be the major product of the reaction.

On his return to Kazan' in 1867, Markovnikov was appointed docent in the Chair of Chemistry. In 1869, he defended his dissertation for the degree of Dr. Chem. [40], "Materials on the question of the mutual influence of atoms in chemical compounds," and in May, 1869, he was appointed Extraordinary Professor of Chemistry, succeeding Butlerov, who had accepted the call to St. Petersburg. Less than a year later, he was promoted to Professor with the Chair of Chemistry. Markovnikov remained at Kazan' until 1872, when he joined six other professors in resigning in protest at the dismissal of the Professor of Physiological Anatomy, Pyotr Frantsevich Lesgaft.[5] Before his resignation could take effect, however, he received the call to the Chair of Chemistry at Odessa University, where he replaced Menshutkin's teacher, Sokolov. The laboratories at Odessa were new and had been well equipped by Sokolov, and he spent a productive time there, despite its brevity. Two years later, he accepted the appointment to the Chair of Chemistry at Moscow University, and he remained at Moscow until he was retired from his Professorship by his opponents, who used the arcane rules of the Department of Education to end his career.

Markovnikov was a prickly individual with strong opinions, and he had no fear in expressing them (his tactless outspokenness led to his ouster from professorships at Kazan' and Moscow). When he felt that the German chemists had attempted to usurp Butlerov's rightful place in the history of the development of structural theory in favor of Kekulé, Markovnikov immediately took them to task in print [41]. Likewise, while a student in Kolbe's laboratory, he did not hesitate to contradict his mentor, whom he admired and respected, when their differences on structural theory came up against Kolbe's intransigence. Leicester [42] and Rocke [39] both recount one story of the interaction between teacher and student thus:

> Naturally Markovnikov, fresh from Butlerov's laboratory, frequently disagreed with Kolbe, but the latter, even when bested in argument, would walk away, saying, "You will see that I am right. Many whom I long ago convinced did not wish to confess it, but now say the same thing. So it will be in this case."
>
> Finally, when Markovnikov was ready to leave Leipzig and was preparing to publish his results, he had a conference with Kolbe. He tells the story as follows: "The heart of the dispute was the old oxygen theory. 'You don't understand me because you are not used to my formulas,' said Kolbe; 'I will express your thoughts in your own formulas.' 'Aha,' I thought, 'now, Herr Professor, you are mine.'... He began to write, stopped halfway

---

[5] Pyotr Frantsevich Lesgaft (Пётр Францевич Лесгафт, 1837–1909) was a teacher, anatomist, physician, and social reformer. He obtained his Dr. Med. at St. Petersburg in 1865, and in 1868 he was called to Kazan' as Professor and Head of the sub-department of Physiological Anatomy. In 1871, he was dismissed from the university for protesting the arbitrary actions of the governing faction of the university administration. He returned to the Medical-Surgical Academy, and led the first group of women students admitted to the institution. He was the founder of the modern system of physical education and medical control in physical training, as well as one of founders of the field of theoretical anatomy. In 1893, he established a Biology laboratory that became, in 1918, the P.F.Lesgaft Institute of Natural Science.

through the formula, thought a minute, then set the pencil down. 'Ja, Sie haben Recht.' Then he completed the formula and said once more, 'Yes, yes, this is true; you are right,' and somewhat confusedly began to explain something. I quickly retired, to spare the self-esteem of an honored teacher. A year later, I received from him a pamphlet on another of our disputed questions. In it he developed his theoretical ideas in detail, but now he wrote the weight of oxygen in the new way."

In 1893, Markovnikov was removed from his Chair, although he was allowed to retain an honorary membership of the University, and to retain his laboratory. The root cause of his removal was the University Statue of 1881, which repealed much of the autonomy that the universities had enjoyed under the University Statute of 1863. Markovnikov's unwillingness to accept the new, strict government supervision meant that his downfall was only a matter of time. The part of the statute used to engineer his ouster was an arcane rule that stated that after 25 years, professors could be subjected to mandatory retirement from their positions. Markovnikov had assumed his first position as Extraordinary Professor at Kazan' in 1868. Still, not every Russian professor was removed from his Chair after 25 years, and, as Leicester points out, the blow to Markovnikov was hard; the blow fell even harder on his student, Aleksei Evgen'evich Chichibabin, as we shall see later. Markovnikov continued to work until the end: he died soon after his return from St. Petersburg, where, in December, 1903, he had presented a review of his work to the Russian Chemical Society.

The first glimmerings of the towering intellect that Markovnikov possessed were provided by his M. Chem. dissertation. This work contained ideas decades ahead of their time, but Markovnikov refused to publish it in full in any language but Russian—despite the urgings of his mentor, Butlerov, and other eminent Russian organic chemists. As a consequence, many of Markovnikov's concepts were attributed to western European chemists who gradually developed the same ideas over the next three decades.

Markovnikov's early work on structural theory had two major thrusts: to explore and elucidate the "reciprocal" influences of atoms in molecules, and to test the limits of structure in organic compounds. As part of the latter thrust, Markovnikov sought to test the belief—widespread at the time—that rings smaller than five members and larger than seven members could not be prepared. He accomplished the first synthesis of a cyclobutane derivative in 1881 [43] by the treatment of β-chloropropionic acid with base to generate 1,3-cyclobutanedicarboxylic acid. In 1882, suberone had been described by Spiegel, but Spiegel did not realize that suberone possesses a seven-membered ring, and so stated in that paper that rings larger than six-membered could not exist [44]. By a sequence of reactions involving reduction of the ketone with sodium in alcohol, and reduction if the alcohol with hydriodic acid, Markovnikov was able to obtain a hydrocarbon that exhibited all the hallmarks of an alkane, and he then went on to establish it as cycloheptane (or, as he called it, heptamethylene) [45].

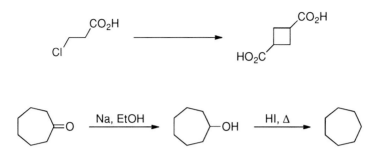

Markovnikov's interests in hydrocarbons continued throughout his life: he wrote many papers on the composition of Caucasus oils, and characterized many of the hydrocarbons in those oils [46]. He was also interested in the chemistry of alicyclic compounds, and especially hydroaromatic compounds, publishing a series of papers on cyclohexane derivatives [47]. Markovnikov also authored a series of papers on pyrotartaric acid (methylsuccinic acid) and its isomers [48].

Despite his other major contributions to the development of modern organic chemistry, Markovnikov has gone into modern organic chemistry textbooks for one reason, and one reason alone: Markovnikov's Rule for addition to alkenes. This rule was first promulgated in 1870 [49] as a four-page addendum to a paper on the isomeric butyric acids, and was followed up by a series of three papers on the formation of chlorohydrins in the *Comptes Rendus* [50], where the rule was explicitly stated:

> En examinant le plupart des cas, suffisamment étudiés, de l'addition directe, je suis arrivé, il y a quelques années, à la conclusion suivante: *Lorsqu'à un hydrocarbure non saturé, renfermant des atomes de carbone inégalement hydrogenés, s'ajoute un acide haloïdhydrique, l'élément électronégatif se fixe sur le carbone le moins hydrogéné...*

Hughes has pointed out that the evidence presented by Markovnikov in his 1870 paper was meager, at best, and he suggests that the rule may actually be the result of an inspired guess [51]. However, Markovnikov's insights into structure and reactivity in organic compounds were highly developed by comparison with his contemporaries, so this judgment may be somewhat harsh. Certainly by 1875, when he published the papers in the *Comptes Rendus*, he had established a solid experimental basis for his rule based on his studies of halohydrin formation and other additions to alkene hydrocarbons.

4.4   Continuing Development at Kazan'

4.4.1   Butlerov's Legacy: The Kazan' School

With Butlerov's call to the chair at St. Petersburg, Kazan' University was faced with the task of replacing him. Although Butlerov delayed his departure in order to allow an orderly transition, the university was faced with a rather unwelcome dilemma. The obvious choice to replace Butlerov as Professor was his student, Markovnikov, who had already substituted for his mentor while Butlerov was abroad defending his place in the development of structural theory. However, as we have seen in the previous section, Markovnikov's fiery nature was not something that the university administration would necessarily relish dealing with. Since there was a second professorship also vacant, the administration chose to fill both positions, with Markovnikov in the *kafedra*. The initial choice fell on Aleksandr Nikiforovich Popov (Александр Никофорович Попов, *ca.* 1840–1881), but before the university could offer him the position, Popov had accepted the call to Warsaw. In some ways, this was fortunate for Kazan': Popov's health was not robust, and he accomplished little more of note after his departure for Warsaw. The university's second choice as Markovnikov's junior colleague was Aleksandr Mikhailovich Zaitsev (Александр Михайлович Зайцев, 1841–1910) [52]. While Zaitsev may have provided a counterpoint to Markovnikov from the university's perspective, the two chemists did not like each other at all, and, in fact, they carried on a life-long feud [53].

4.4.2 Markovnikov and Zaitsev: Two Faces of the Same Coin

Contemporaries, Zaitsev and Markovnikov had both studied under Butlerov. As we saw in the previous section, however, much of Markovnikov's career was passed at Moscow University, where he built the chemistry department into an important center for organic chemistry. Zaitsev, on the other hand, remained at Kazan', and achieved a forty-year career in the *kafedra* of organic chemistry.

Despite his outspoken and argumentative nature, Markovnikov's career was surprisingly conventional, and followed the traditional sequence of events: he completed his *kandidat* degree, then worked as an assistant before traveling abroad on a *komandirovka* to western Europe. He then returned to his position at the university, wrote up his dissertation for the degree of M. Chem., and was promoted to Extraordinary Professor. After some additional work in Russia, he wrote and defended his dissertation for the degree of Dr. Chem., and was promoted to Ordinary Professor. As we have seen, his lack of tact and inability to compromise with administrators and his more conservative colleagues eventually led to his forced retirement at Moscow: not even his international stature could save him.

By contrast, Markovnikov's contemporary, Zaitsev, began his career with several ill-advised, impulsive moves that nearly ended his academic career before it had begun. However, the fact that he was able to recover from his early mistakes to occupy the Chair of Chemistry for four decades speaks highly of his adroitness. Where Markovnikov was a political *naif*, Zaitsev's career reflects a high level of political skill—knowing what to do, and when. Zaitsev flouted tradition, but he always seemed to land on his feet.

So why describe the two students as two faces of the same coin? The beginnings to their careers certainly contrasted: Markovnikov was the observant traditionalist, Zaitsev the iconoclastic opportunist. Their characters, likewise, contrast: Markovnikov was the demonstrative hothead, Zaitsev the cool customer; and yet, both men inspired real loyalty among their students. Perhaps most important for the development of organic chemistry in Russia, Markovnikov was a brilliant theoretician, while Zaitsev was a superb experimentalist. Their complementary skills and strengths worked in harmony (even if they did not) to advance the cause of Russian organic chemistry.

4.4.3 Aleksandr Mikhailovich Zaitsev

Александр Михайлович
Зайцев (1841-1910)
Aleksandr Mikhailovich
Zaitsev (Saytzeff)

Zaitsev was the son of Mikhail Savvich Zaitsev and his second wife, Nataliya Vasil'evna Lyapunov, the sister of astronomer Mikhail Vasil'evich Lyapunov (1820–1868). Zaitsev's family had been in Kazan' since the time of Ivan the Terrible, (1530–1584), who had conquered the Tatar Khanate of Kazan' in 1552. Zaitsev's family was prominent in the local guilds, which conferred a certain stature on the family (his grandfather, Savva Stepanovich Zaitsev, was an elder in the cathedral, and a member of the group overseeing its renovation). Zaitsev's father had destined him for a career in the guilds, but, at the urging of his brother-in-law, Zaitsev was permitted to enter Kazan' University. His father insisted, however, that he enter the Juridicial-Economic faculty, which required Latin; the Gymnasium did not teach Latin, so his uncle, Lyapunov, personally taught him the Latin he needed for the entrance examination. Zaitsev entered Kazan' University in 1858 as a student in the cameral section, which meant that he came under the influence of Butlerov.

It us unclear at what time Zaitsev became a truly committed disciple of Butlerov. He graduated with his *Diplom* in economic science in 1862. Shortly before his graduation, his father had died, and the family business had been sold, with the proceeds being split among the sons. Zaitsev took his share of the proceeds and immediately followed his older brother, Konstantin Nikolaevich,[6] to the Marburg laboratory of Hermann Kolbe. This was a risky move—without the degree of *kandidat*, Zaitsev was not eligible to occupy a salaried place at a Russian university. Nevertheless, Zaitsev spent three semesters with Kolbe from 1862 to 1864, and then he spent a year with Charles Adolphe Wurtz in Paris before returning to Marburg for a final semester with Kolbe. After this semester, Kolbe was called to the Chair at Leipzig, but Zaitsev had run out of funds by then, so he returned to Russia instead of following Kolbe to his new laboratory (although he is actually listed first among the *praktikanten* at Leipzig [54]).

While at Marburg, Zaitsev realized that his departure from Kazan' without the *kandidat* had placed his future in jeopardy. To rectify the situation, he submitted a 75-page, hand-written dissertation [55] discussing Kolbe's views on the rational constitution of chemical compounds to Kazan'. The move was not well considered: this dissertation, which promoted the views of Kolbe, structural theory's staunchest opponent, was submitted to examination by Butlerov, its most ardent proponent. Butlerov's usual equanimity was not apparent in his review of this dissertation, which he subjected to unusually harsh criticism. He pointed out inconsistency after inconsistency, characterizing it as "a poor rendering of the German." The degree was not awarded, and, given the circumstances, one might be justified in assuming that this was the end of Zaitsev's career. It was not.

While in western Europe, Zaitsev had accomplished a significant amount of research, and had published half a dozen papers in the *Annalen*, the *Comptes Rendus*, and the *Bulletin de la Société Chimique de Paris* [56]. This may have been the basis on which Butlerov decided to facilitate his return to Kazan' in 1865: despite his impetuosity in flouting Russian tradition, someone this productive could obviously make important contributions to organic chemistry in Russia. Whatever the reason, Zaitsev returned to Kazan' as an unpaid assistant (a "private person") in Butlerov's laboratory, and proceeded to write up his work on diaminosalicylic acid for the degree of *kandidat* [57]. He was immediately appointed as the Assistant in charge of the agronomy laboratory.

In order to become a Professor of Chemistry, Zaitsev needed the M. Chem. degree, which was awarded by the Physics-mathematics faculty, and his graduation with a degree in *cameral* science did not qualify him to pursue this degree. Again, Zaitsev's ability to find a way around roadblocks came to his rescue: finding that a doctoral degree from a foreign university would officially satisfy the

---

[6] Konstantin Nikolaevich Zaitsev (Константин Николаевич Зайцев (b. 1840) was the first of the Zaitsev brothers to study chemistry at Kazan', and was the first of a series of Butlerov students to study with Kolbe. He returned to Russia in 1863, after two semesters with Kolbe, and was appointed as a lecturer in analytical chemistry. In 1868, he was invited by the owners of the Krestovnikov soap factory, the major chemical factory in Kazan', to join the staff of the plant to organize the laboratory there. In 1872, he was appointed director of the laboratory, position he held until 1907.

prerequisites to allow him to submit for the M. Chem. degree, Zaitsev submitted a Ph.D. dissertation on sulfoxides [58] to Kolbe at Leipzig. Although the degree was awarded, it still took Butlerov's intervention before Zaitsev (or Markovnikov, for that matter) was permitted to pursue the degree of M. Chem. Zaitsev then proceeded to write the same work up for his M. Chem. degree [59]; he graduated into the degree in 1868, and was promoted to Extraordinary Professor of Chemistry in 1869. A year later, he submitted a dissertation on the reduction of fatty acids [60] for the degree of Dr. Chem. Markovnikov was appointed as the primary examiner of this dissertation, and he wrote an overtly positive review that was meant to be read between the lines. Fortunately for Zaitsev, Butlerov, whose opinion carried great weight with the faculty, knew of the sour relationship between his two students, and his positive recommendation led to Zaitsev being awarded the degree, and being promoted to Ordinary Professor (albeit on a 19-12 split vote). Thus began a career that spanned another four decades, and that inspired strong loyalty among his students (as evidenced by the presentations made to him by his students on the occasions of his various anniversaries, Fig. 4.3).

The contrast between Zaitsev and Markovnikov becomes apparent when one examines their research output. From the first, it was evident that Markovnikov was a brilliant theoretician, and it was here that his major contributions lay. Even his experimental work was aimed at serving the cause of structural theory. Zaitsev, on the other hand, was an experimentalist of the first rank, and all the 120 research publications listed in his biography [52c] deal with synthetic organic reactions. Like his contemporaries, Zaitsev routinely followed the publication of his work in Russian by publication of the same work in German (usually in the *Annalen* or the *Journal für Praktische Chemie*).

Zaitsev's early synthetic contributions came while he was a student in Kolbe's laboratory. While in Marburg, he carried out the reduction of dinitrosalicylic acid with hydrogen iodide to form the diamino compound. Like all very electron-rich aromatics, this compound is very susceptible to oxidation by air or mild oxidizing agents, and Zaitsev stabilized it in the usual manner: he converted the free amine into the corresponding salt [57a,b], which presumably exists as the zwitterion, as well, and is therefore much less electron-rich, and much less susceptible to oxidation.[7] This work with easily oxidized compounds reveals Zaitsev's skill as a synthetic chemist, a skill he would again show in his independent career.

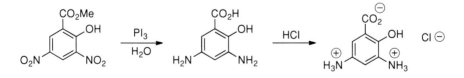

---

[7] In these papers, Zaitsev reports that the colorless acid becomes a black mass, and that all attempts at preparing a nitrate salt—using an oxidizing acid—also gave a black mass.

Fig. 4.3 Albums presented to Zaitsev on the occasions of his anniversaries

His study of the reactions of chloroacetic ester with potassium cyanate under reflux in ethanol, carried out partly in Wurtz' Paris laboratory and partly in Kolbe's laboratory, was reported in 1865 [56c-e]. Unlike the analogous reaction with potassium cyanide, Zaitsev found that the reaction with potassium cyanate was complicated; a modern reading of the paper shows that under these conditions, cyanate anion acts as an oxygen nucleophile, but that the initial product then undergoes subsequent addition to give an isourea derivative of glycolic acid, which was obtained on acid hydrolysis of the initial product. In the same paper, Zaitsev showed that the potassium cyanate also reacts with the solvent to give the major product of the reaction.

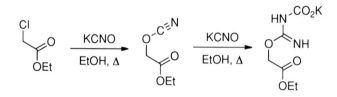

While at Marburg, Zaitsev studied the chemistry of organosulfur compounds. As part of this work, he discovered the oxidation of dialkyl sulfides to the corresponding sulfoxides [61], thus identifying a new class of organic sulfur compounds. At the same time, he treated amyl ethyl sulfide with methyl iodide to obtain the corresponding sulfonium salt [62], another new class of organosulfur compounds.

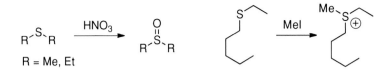

Throughout his career, Zaitsev studied redox reactions, although this did not form a major part of his research output. His work for the Dr. Chem. degree involved the reduction of acid chlorides with sodium amalgam in ether; Zaitsev found that it was important to buffer the reduction with acetic acid, since the reaction slowed or stopped as the reaction mixture became more basic (presumably because of the formation of the ester, which was much more resistant to oxidation). Using this method, Zaitsev described the first synthesis of γ-butyrolactone by reduction of succinyl chloride [63].

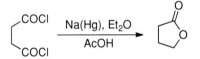

Once he began his independent career, Zaitsev continued his mentor, Butlerov's, line of research with organozinc reagents in organic synthesis [64]. Butlerov had prepared *tert*-butyl alcohol from the reaction between dialkylzinc reagents and acid chlorides (notably, phosgene!), and Zaitsev's first major independent contribution was to extend the range of the reaction by substituting an alkyl iodide—especially reactive allyl iodides—and metallic zinc for the pyrophoric dialkylzinc reagent [65]. Over the course of several decades, Zaitsev and his students (who included his own brother, Mikhail Mikhailovich[8]) explored the addition of alkylzinc iodides to a variety of carbonyl partners, including ketones [66], formate esters [67], and anhydrides [68], in addition to acid chlorides. It is interesting to note that Mikhail Zaitsev observed that the reaction of an alkylzinc iodide with an anhydride could be used to produce a ketone. The observation that the attempted addition of propylzinc iodide to 4-heptanone (butyrone) gave not the expected

---

[8] Mikhail Mikhailovich Zaitsev (Михаил Михайлович Заицев, 1845–1904) was the younger brother of Zaitsev, and like his brother he studied under Butlerov. As a student of his brother, he studied the reactions of organozinc nucleophiles, but he also studied the reactions of hydrogen with unsaturated alcohols in the presence of palladium and platinum black [Saytzeff (1873) Ueber die Einwirkungdes vom Palladium absorbtieren Wasserstoffes auf einige organische Verbindungen. J Prakt Chem 6:128-135]; this work led to the first industraial hydrogenation of fats in Russia by Fokin in 1909–1910.

tertiary alcohol, but 4-heptanol instead [69] may be the first observation of the now well-known β-hydride transfer from carbon during organometallic additions to carbonyl compounds. It is worthwhile noting that the allylmetal is much less prone to β-hydrogen transfer reactions.

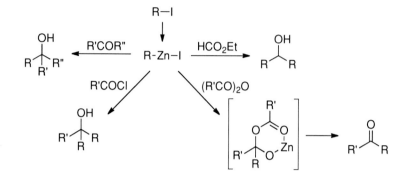

The versatility of organozinc nucleophiles is nicely illustrated by transformations that came out of Zaitsev's Kazan' laboratory.

Later in his career, Zaitsev began to investigate oxidation reactions. As part of this effort, he undertook the oxidation of oleic acid and its geometric isomer, elaidic acid with potassium permanganate [70]. The choice of substrate for this reaction may not have been random, since Konstantin Zaitsev was, at the time, director of the chemistry laboratory at the Krestivnikov Brothers' chemical plant, where high-grade glycerine for use in cosmetics and explosives was manufactured. The process produced considerable amounts of soap, and Zaitsev's interest may have been in developing new products from the by-product of glycerine manufacture.

The oxidation provided the dihydroxystearic acids, but Zaitsev did not pursue the reaction further, and it was his student, Wagner, who converted the reaction into a useful synthetic method.

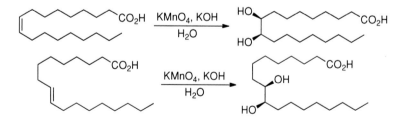

References

1. Leicester HM (1947). The history of chemistry in Russia prior to 1900. J Chem Educ 24: 438-443.
2. Biographies: (a) Morachevskii AG (2007). Nikolai Aleksandrovich Menshutkin (to 100th anniversary of his death). Russ J Appl Chem 80:166-171 [original Russian text: Zh Prikl Khim 80:167-172]; (b) Menschutkin B (1907). Nikolai Alexandrowitsch Menschutkin. Ber Dtsch Chem Ges 40:5087-5098; (c) Lutz-riga O (1907). Nikolai Menschutkin. Z Angew Chem 20:609-610.
3. Menshutkin's obituary of Sokolov: Menshutkin NA (1878). Zh Russ Fiz-Khim O-va 10:8-15.
4. (a) Menshutkin NA (1866). O vodorode fosforistoi kisloty, ne sposobnom k metallicheskoi zameshcheniyu pri obyknovennykh usloviyakh dlya kislot [On the hydrogen of phosphorous acid, not capable of being replaced by metals in the normal way for acids]. M Chem Diss, St. Petersburg; (b) Menschutkin N (1865). Ueber die Einwirkung des Chloracetyls auf phosphorige Säure. Justus Liebigs Ann Chem 133:317-320.
5. Menshutkin NA (1869). Syntez i svoitsva ureidov [Synthesis and properties of ureides]. Dr Chem Diss, St. Petersburg.
6. Menschutkin N (1882). Ueber die Zersetzung des tertiären Amylacetats durch Wärme. Ber Dtsch Chem Ges 15:2512-2518.
7. (a) Menschutkin N (1883). Sur la détermination de l'isomérie des acides par la vitesse initiale de leur éthérification. Rec Trav Chim Pays-Bas 2:117-120; (b) Menschutkin N (1878). Aetherification primärer Alkohole. Ber Dtsch Chem Ges 11:1507-1511; Ueber die Aetherification der secundären Alkohole. ibid 11:2117-2120; (c) Menschutkin N (1879). Ueber den Einfluss der Isomerie der Alkohole und der Säuren auf die Bildung zusammengesetzter Aether. Justus Liebigs Ann Chem 197:193-225; (d) Menschutkin N (1882). Anleitung zur Bestimmung der Isomerie der Alkohole und Säuren mit Hülfe ihrer Aetherificirungsdaten. J Prakt Chem 26:103-120; (e) Menschutkin N (1879). Ueber den Einfluß der Isomerie der Alkohole und der Säuren auf die Bildung zusammengesetzter Aether. Justus Liebigs Ann Chem 195:334-364.
8. (a) Menshutkin NA (1890). O vlyanii khimicheskoi nedeyatel'noi zhidnoi sredy na skorost' soedineniya trietilamina s iodgidrinami [On the influence of chemically inactive liquid media on the speed of formation of the compound of triethyla mine with iodohydrine]. Zh Russ Fiz-Khim O-va 22:393-409 (b) Menschutkin N (1890). Beiträgen zur Kenntnis der Affinitätskoeffizienter der Alkylhaloide und der organischen Amine. Z Physik Chem 5:589-600; (1890). Über die Affinitätskoeffizienten der Alkylhaloide und der Amine ibid 6:41-57.
9. Menschutkin NA (1895). Vyanie chisla tsepei na skorost' obrazovaniya aminov [The influence of the number of chains on the rate of formation of amines]. Zh Russ Fiz-Khim O-va 27: 96-118, 137-157.
10. Menschutkin N (1905). Ueber den Einfluss indifferenter Lösungsmittel bei der Alkylirung organischer Basen. Ber Dtsch Chem Ges 38:2465-2466.
11. Hughes ED, Ingold CK (1935). Mechanism of substitution at a saturated carbon atom. Part IV. A discussion of constitutional and solvent effects on the mechanism, kinetics, velocity, and orientation of substitution. J Chem Soc 244-255.
12. For biographies of Borodin, see: (a) Kauffman GB, Bumpass K (1988). An Apparent Conflict between Art and Science: The Case of Aleksandr Porfir'evich Borodin (1833-1887). Leonardo 21:429-436. (b) Podlech J (2010). "Try and fall sick…"—The Composer, Chemist, and Surgeon Aleksandr Borodin. Angew Chem Int Ed 49:6490-6495. For a biography of Borodin largely focused on his musical accomplishments, see (c) Abraham G, Lloyd-Jones D (1986). Alexander Borodin. In: Brown D, Abraham GE, (eds) The New Grove Russian Masters 1. W.W. Norton & Company, New York. pp. 45-76.
13. Gordin MD (2006). Facing the music: How original was Borodin's chemistry? J Chem Educ 83:561-565.

14. Rae ID (1989). The research in organic chemistry of Aleksandr Borodin (1833-1887). Ambix 36:121-137.
15. Stassow W (1993). Meine Freunde Alexander Borodin und Modest Mussorgsky. Ernst Kuhn, Berlin, p. 138.
16. Shine HJ (1989). Borodin and the benzidine rearrangement. J Chem Educ 66:793-794.
17. (a) Borodin A (1861). Sur les dérivés monobromés des acides valérique et butyrique. Bull Soc chim Paris 252-254; (b) Borodine A (1861). Ueber Bromvaleriansäure und Brombuttersäure. Justus Liebigs Ann Chem 119:121-123; (c) Borodine A (1861). Ueber die Monobrombaldriansaure und Monobrombuttersaure. Z Chem Pharm 4:5-7; (1869). Ueber die Produkte von Bromdampf auf Silbersalze: Öl- und Baldriansäure. ibid 12:342; (d) Borodin AP (1869). O produktakh deitsviya parov broma na serebryanyya soli kislot: maslyanoi i valer'yanovoi [On the product of bromine vapor with silver salts of acids: oleic and valerian] Zh Russ Fiz-Khim O-va 1:31-32.
18. (a) Hunsdiecker C, Hunsdiecker H, Vogt E (1939). Method of manufacturing organic chlorine and bromine derivatives. U.S. Patent No. 2176181.
19. (a) Hunsdiecker H, Hunsdiecker C (1942). Über den Abbau der Salze aliphatischer Säuren durch Brom. Ber Dtsch Chem Ges 75:291-297. (b) Johnson RG, Ingham RK (1956). The degradation of carboxylic acid salts by means of halogen – The Hunsdiecker reaction. Chem Rev 56:219–269. (c) Wilson CV (1957). The reaction of halogens with silver salts of carboxylic acids. Org React 9:332-387.
20. (a) Borodin A (1862). Fait pour servir a l'histoire des fluorures et préparation du fluorure de benzoyle. Comptes Rend. Acad Sci 55:553-556; (b) Borodine A (1863). Zur Geschichte der Fluorverbindungen und über das Fluorbenzoyl. Justus Liebigs Ann Chem 126:58-62.
21. (a) Wurtz A (1872). Sur un aldéhyde-alcool. Bull Soc chim Paris 17:436-442; (b) Wurtz A (1872). Ueber einen Aldehyd-Alkohol. J Prakt Chem 5:457–464; (c) Wurtz A (1872). Sur un aldéhyde-alcool. Compt Rend Acad Sci 74:1361-1367.
22. (a) Borodin A (1864). Ueber die Einwirkung des Natriums auf Valeraldehyd. J Prakt Chem 93:413-425; Ueber die Einwirkung des Natriums auf Valeraldehyd. Z Chem Pharm 7: 353-364.
23. (a) Kekulé A (1869). Condensationsproducte des Aldehyds; — Crotonaldehyd. Ber Dtsch Chem Ges 2:365-368; (1870). Ueber die Condensation der Aldehyde. ibid 3:135-137.
24. Gordin MD (2004). A Well-Ordered Thing: Dmitrii Mendeleev and the Shadow of the Periodic Table. Basic Books, New York.
25. Gordin MD (2006). Beilstein Unbound: The Pedagogical Unraveling of a Man and his Handbuch. In: Kaiser D (ed.) Pedagogy and Practice of Science. MIT Press, Cambridge; Ch. 1, pp. 11-39.
26. Beilstein F (1856). Ueber die Diffusion von Flüssigkeiten. Justus Liebigs Ann Chem 99:165-197.
27. Beilstein F (1858). Ueber das Murexid. Justus Liebigs Ann Chem 107:176-191.
28. Beilstein F (1860). Ueber die Identität des Aethylidenchlorürs und des Chlorürs des gechlorten Aethyls. Justus Liebigs Ann Chem 113:110-112.
29. For lucid account of this election and the public outrage it caused, see Ref 23, ch. 5.
30. Beilstein F, Geitner P (1866). Untersuchungen über Isomerie in der Benzoëreihe. Sechste Abhandlung. Ueber das Verhalten der Homologen des Benzols gegen Chlor. Justus Liebigs Ann Chem 139:331-342.
31. (a) Gomberg M (1900). An instance of trivalent carbon: triphenylmethyl. J Am Chem Soc 22:757-771. (b) Gomberg M (1901). On trivalent carbon. ibid 23:496-502.
32. Kharasch MS, Mayo FR (1933). The peroxide effect in the addition of reagents to unsaturated compounds. I. The addition of hydrogen bromide to allyl bromide. J Am Chem Soc 33: 2468-2496.
33. Hey DH, Waters WA (1937). Some organic reactions involving the occurrence of free radicals in solution. Chem Rev 21:169-208.
34. Beilstein F (1872). Ueber den Nachweis von Chlor, Brom und Jod in organischen Substanzen. Ber Dtsch Chem Ges 5:620-621.

35. For biographies of Markovnikov in Russian, see: (a) Dem'yanov NYa (1904). Pamyati Vladimira Vasil'evicha Markovnikova [Memories of Vladimir Vasil'evich Markovnikov]. Zh Russ Fiz-Khim O-va 36:345. (b) Yakovkin AA (1904). Pamyati Vladimira Vasil'evicha Markovnikova [Memories of Vladimir Vasil'evich Markovnikov]. Zh Russ Fiz-Khim O-va 36:181. (c) Beketov NN (1904). Pamyati Vladimira Vasil'evicha Markovnikova [Memories of Vladimir Vasil'evich Markovnikov]. Zh Russ Fiz-Khim O-va 36:180. (d) Kablukov IA (1905) Vladimir Vasil'evich Markovnikov, M., 1905. Zh Russ Fiz-Khim O-va 37:247-303. (e) Platé AF, Bykov, GV, Eventova MS. Vladimir Vasilevich Markovnikov. Ocherk zhizni i deyatelnosti. 1837–1904 [Vladimir Vasilevich Markovnikov. An account of his life and work. 1837-1904]. Moscow, 1962. Biographical information in western languages may be found in: (f) Decker H (1906). Wladimir Wasiliewitsch Markownikoff. Ber Dtsch Chem Ges 38:4249-4259. (g) Mills EJ (1905). Wladimir Wasiljewitsch Markownikoff. J Chem Soc 87:597-600. (h) Leicester HM (1941). Vladimir Vasil'evich Markovnikov. J Chem Educ 18:53-57. (i) Leicester HM (1966). Kekulé, Butlerov, Markovnikov. Controversies on Chemical Structure From 1860 to 1870. In Kekulé Centennial, Advances in Chemistry Series, Washington, DC, 13-23. (j) Bykov GV (2008). Markovnikov, Vladimir Vasilevich. Complete Dictionary of Scientific Biography.
36. Markovnikov VV (1860). O al'degidakh i ikh otnosheniyakh k alkogol'am i ketonam [On aldehydes and their relationship to alcohols and ketones]. Kand Diss, Kazan'.
37. Markovnikov V (1865). Ob izomerii organicheskikh soedinenii [On the isomerism of organic compounds]. M Chem Diss, Kazan'.
38. Quoted in Arbuzov AE (1948). Kratkii ocherk razvitiya organicheskoi khimii v Rossii [A brief account of the development of organic chemistry in Russia]. Moskva, Izd-vo Akademii nauk SSSR. p. 113, which cites Pamyati A.M. Butlerova [Letters of A. M. Butlerov] (1887) Zh Russ Fiz-Khim O-va 19:87.
39. Rocke AJ (1993). The Quiet Revolution. Hermann Kolbe and the Science of Organic Chemistry. University of California Press, Berkeley. p. 316.
40. Markovnikov V (1869). Materialy voprosu o vzaimnom vliyanii atomov v khimicheskikh soedineniyakh [Materials on the question of the mutual influence of atoms in chemical compounds]. Dr Chem Diss, Kazan'.
41. Markownikoff W (1865.) Zur Geschichte der Lehre uber die chemische Structur. Z Chem nf 1:280-287.
42. Leicester HM (1941). Vladimir Vasil'evich Markovnikov. J Chem Educ 18:53-57; the quotation is taken from pp. 54-55.
43. Markownikoff W, Krestownikoff A (1881). Aus dem chemischen Universitätslaboratorium Moscau von Prof. Markownikoff. Tetrylendicarbonsäure (Homo). Justus Liebigs Ann Chem 208:333-349.
44. Spiegel, A. (1882). Ueber das Suberon. Justus Liebigs Ann Chem 211:117-120.
45. (a) Markownikoff W (1890). Derivés de l'heptaméthylène. Comptes Rend. 110:466-468. (b) Markovnikoff, W. (1892). Sur un nouvel hydrocarbure, le subérène. Comptes Rend. 115:462-464.
46. (a) Markovnikoff W, Oglobine W (1884). Recerches sur le pétrole du Caucase. Ann Chim Phys, Ser 6 2:372-484. (b) Markownikoff W (1886). Die aromatischen Kohlenwasserstoffe des kaukasischen Erdöls. Justus Liebigs Ann Chem 234:89-115. (c) Markownikoff W, Spady J (1887). Zur Constitution der Kohlenwasserstoffe, C_nH_{2n}, des kaukasischen Petroleums. Ber Dtsch Chem Ges 20:1850-1853. (d) Markownikoff W (1890). Ueber das Rosenöl. Ber Dtsch Chem Ges 23:3191. (e) Markownikoff W, Reformatsky A (1893). Untersuchung des bulgarischen Rosenöls. J Prakt Chem 48:293-314. (f) Markownikoff W. (1892) Die Naphtene und deren Derivate in dem allgemeinen System der organischen Verbindungen. (Erster Teil). J Prakt Chem 45:561-580; Die Naphtene und deren Derivate in dem allgemeinen System der organischen Verbindungen. (Zweiter Theil.) ibid 46:86-106.

47. For some leading references, see: Markownikoff W (1892). Zur Geschichte der Hydrobenzoësäuren. Ber Dtsch Chem Ges 25:370-372; (1892). Ueber die Heptanaphtensäure (Hexahydrobenzoësäure). ibid 25:3355; (1894). Ueber die isomeren Octonaphtensäuren (Cyclohexanmethylcarbonsäure). J Prakt Chem 49:64-89.

48. Markownikoff W (1872). I. Ueber die isomeren Pyroweinsäuren. Justus Liebigs Ann Chem 182:324-346; (1873). Die dritte isomere Pyroweinsäure. Ber Dtsch Chem Ges 6:1440; (1876). Ueber die normale Pyroweinsäure. ibid 9:787-788.

49. Markownikoff W (1870). I. Ueber die Abhängigkeit der verschiedenen Vertretbarkeit des Radicalwasserstoffs in den isomeren Buttersäuren. Justus Liebigs Ann Chem 153:228-259.

50. Markovnikoff W (1875). Sur les loies qui régissent les réactions de l'addition directe. Comptes Rend. 81:668-671, 728-730, 776-779.

51. Hughes P (2006). Was Markovnikov's Rule an Inspired Guess? J Chem Educ 83:1152-1154.

52. For biographies of Zaitsev in Russian, see: (a) Chugaev LA (1910). Pamyati A. M. Zaitseva i Kannitsaro [Memories of A. M. Zaitsev and Cannizzarro]. Zh Russ Fiz-Khim O-va 42:1318. (b) Reformatskii AN (1911). Biografiya Aleksandr Mikhailovicha Zaitseva [Biography of Aleksandr Mikhailovich Zaitsev]. Zh Russ Fiz-Khim O-va 43:876, abstr. 6. (c) Klyuchevich AS, Bykov GB (1980). Aleksandr Mikhailovich Zaitsev. Izdatel'stvo "Nauka", Moscow. (d) Reformatskii SN (1910). Pamyati Prof. A. M. Zaitseva [Memories of Prof. A. M. Zaitsev]. Univ Izv Kiev, No 11, Pril 2 In Protokol Fiz-Khim O-va, p 1. (e) Arbuzov AE (1948). Kratkii ocherk razvitiya organicheskoi khimii v Rossii [A brief account of the development of organic chemistry in Russia]. USSR Academy of Science Publishers, Moscow. For biographical information in western languages, see: (f) (1910). Alexander Micholajeff Saytzeff. Ber Dtsch Chem Ges 43:2784. (g) Lewis DE (1995). Aleksandr Mikhailovich Zaitsev (1841-1910). Markovnikov's conservative contemporary. Bull. Hist. Chem 17/18:21-30. (h) Lewis DE (2011). A. M. Zaitsev: Lasting contributions of a synthetic virtuoso a century after his death. Angew Chem Int Ed 50:6452-6458; A. M. Saytzeff: bleibendes Vermächtnis eines Virtuosen der Synthesechemie. Angew Chem 123:6580-6586.

53. Lewis DE (2010). Feuding Rule-Makers: Vladimir Vasil'evich Markovnikov (1838-1904) and Aleksandr Mikhailovich Zaitsev (1841-1910). A Commentary on the Origins of Zaitsev's Rule. Bull. Hist. Chem. 35:115-124.

54. See Ref. 39 pp. 123, 284.

55. Zaitsev AM (1863). Teoreticheskie vzglyady Kol'be na ratsional'nuyu konstitutsiyu organicheskikh soedinenii i ikh svyaz' s neorganicheskimi [The theoretical views of Kolbe on the rational constitution of organic compounds and their relationship with inorganic (compounds-DEL)]. This dissertation is referred to in ref. 45(c), and may have been one contributing factor to the animus between Zaitsev and Markovnikov—see ref. 53.

56. (a) Saytzeff A (1865). Ueber Diamidosalicylsäure. Justus Liebigs Ann Chem 133:321-329. (b) Saytzeff A (1865). Sur l'acide diamidosalicylique et sur ses combinaisons avec les acides. Bull Soc Chim Paris 3:244-250. (c) Saytzeff A (1865). Ueber die Einwirkung von cyansäurem Kali auf Monochloroessigsäther. Justus Liebigs Ann Chem 133:329-355; ibid 135:229-236. (d) Saytzeff A (1865). Action du cyanate de potasse sur l'éther monochloroacétique. Comptes Rend Acad Sci 60:671-673. (e) Saytzeff A (1865). Action du cyanate de potasse sur l'éther monochloroacétique. Bull Soc Chim Paris 3:350-356.

57. Zaitsev AM (1866). O diamidosalitsilovnoi kislote [Concerning diamidosalicylic acid]. Kand Diss, Kazan University.

58. Saytzeff A (1866). Über eine neue Reihe organischer Schwefelverbindungen. PhD Diss, Leipzig University.

59. Zaitsev AM (1867). O deistvii azotnoi kisloty na nekotorye organicheskie soedineniya dvuatomnoi sery i o novom ryade organicheskikh sernistykh soedinenii poluchennom pri etoi reaktsii prolozhenie [On the action of nitric acid on certain organic compounds of divalent sulfur and on a new class of organic sulfur compounds obtained from this reaction]. M Chem Diss, Kazan' University. The abstract of this dissertation also appeared in Student Writings of Kazan University, 1867, 3:2; Prilozhenie: Iz khimicheskoi laboratorii Kazanskogo

universiteta. s. 1-24 [Appendix: From the chemical laboratories of Kazan' University. p. 1-24].

60. Zaitsev AM (1870). Novye sposob prevashcheniya zhirnykh kislot v sootvetstvuyushchie im alkogoli. Normal'nyi butil'nyi alkogol' (propilkarbinol) i ego prevashchenie vo vtorichnyi butil'nyi alkogol' (mefil-efil-karbinol) [A new method for converting a fatty acid into its corresponding alcohol. Normal butyl alcohol (propyl carbinol) and its conversion to secondary butyl alcohol (methyl-ethyl-carbinol)], Dr Chem Diss, Kazan' University. The work was also published under the same title: Zaitsev AM (1870) Zh Russ Fiz-Khim O-va 2:292-310.

61. (a) Saytzeff A (1866). XLIII. Ueber eine neue Reihe organischer Schwefelverbindungen. Justus Liebigs Ann Chem 139:354-364; (1867). Ueber die Einwirkung von Salpetersäure auf Schwefelmethyl und Schwefeläthyl. ibid 144:148-156.

62. Saytzeff A (1867). Ueber die Einwirkung von Jodmethyl auf Schwefelamyläthyl. Justus Liebigs Ann Chem 144:145-148.

63. Saytzeff A (1874). 4. Ueber die Reduction des Succinylchlorids. Justus Liebigs Ann Chem 171:258-290.

64. Lewis DE (2002). The beginnings of synthetic organic chemistry: Zinc alkyls and the Kazan' school. Bull Hist Chem 27:37-42.

65. A selection from the 37 papers on the synthesis of alcohols using zinc metal and alkyl iodides includes: (a) Wagner G, Saytzeff A (1875). Synthese des Diäthylcarbinols, eines neuen Isomeren des Amylalkohols. Justus Liebigs Ann Chem 175:351-374. (b) Vagner E, Zaitsev A (1874). Sintez dietilkarbinola, novogo izomera amil'nogo alkogolya [The synthesis of diethylcarbinol, a new isomer of amyl alcohol]. Zh Russ Fiz-Khim O-va 6:290-308. (c) Kanonnikoff J, Saytzeff A (1875). Neue Synthese des secundären Butylalkohols. Justus Liebigs Ann Chem 175:374-378. (d) Kanonnikov I, Zaitsev AM (1874). Novyi sintez vtorichnogo butil'nogo alkogolya [A new synthesis of secondary butyl alcohol]. Zh Russ Fiz-Khim O-va 6:308-312. (e) Kanonnikoff I, Saytzeff A (1877). Ueber Einwirkung eines Gemische von Jodallyl mit Jodäthyl und Zink auf das ameisensaure Aethyl. Justus Liebigs Ann Chem 185, 148-150. (f) Kanonnikov I, Zaitsev AM (1876). O deistvii smesi iodistogo allila s iodistym etilom i tsinka na murav'inyi etil'nyi efir [On the action of a mixture of allyl iodide with ethyl iodide and zinc on formic acid ethyl ester]. Zh Russ Khim O-va Fiz O-va 8:359-363. (g) Saytzeff M, Saytzeff A (1877). Synthese des Allyldimethylcarbinols. Justus Liebigs Ann Chem 185:151-169. (h) Saytzeff A (1877). Bemerkung über Bildung und Eigenschaften der in den vorhergehenden Abhandlungen beschriebenen ungesättigten Alkohole. Justus Liebigs Ann Chem 185:175-183; (1885). Synthese der tertiären gesättigten Alkohole aus den Ketonen; Vorläufige Mittheilung. J Prakt Chem 31:319-320. (1875).

66. (a) Saytzeff M, Saytzeff A (1877). Synthese des Allyldimethylcarbinols. Ann Chem Pharm 185:151-169. (b) Saytzeff A (1885). Synthese der tertiären gesättigten Alkohole aus den Ketonen; Vorläufige Mittheilung, J Prakt Chem 31:319-320. (c) Saytzeff A (1877). Bemerkung über Bildung und Eigenschaften der in den vorhergehenden Abhandlungen beschriebenen ungesättigten Alkohole. Ann Chem Pharm 185:175-183.

67. (a) Wagner G, Saytzeff A (1875). Synthese des Diäthylcarbinols, eines neuen Isomeren des Amylalkohols, Ann Chem Pharm 175:351-374. (b) Kanonnikoff J, Saytzeff A (1875). Neue Synthese des secundären Butylalkohols. Ann Chem Pharm 175:374-378. (c) Kanonnikoff J, Saytzeff A (1877). Ueber Einwirkung eines Gemisches von Jodallyl mit Jodäthyl und Zink auf das ameisensaure Aethyl. Ann Chem Pharm 185:148-150.

68. (a) Zaitsev M (1870). O deistvii tsink-natriya na smes' iodistogo efila ili mefila s uksusnym angidridom [On the action of zinc-sodium on a mixture of ethyl or methyl iodide with acetic anhydride.] Zh Russ Fiz-Khim O-va 2:49-51. (b) Kanonnikoff J, Saytzeff M (1877). Zur Darstellung des Jodallyls und des Essigsäureanhydrids. Ann Chem Pharm 185:191-192. (c) Saytzeff A (1907). Über die Einwirkung von Jodzinkallyl auf Anhydride einbasischer Säuren. J Prakt Chem 76:98-104; this paper acknowledges the following students: F Petroff, N Musuroff, S Chowansky, G Andreeff, B Chonowsky, and A Lunjack.

69. Ustinoff D, Saytzeff A (1886). Über die Einwirkung von Jodpropyl und Zink auf Butyron. Bildungsweise des Dipropylcarbinols, J Prakt Chem [ii] 34:468-472.
70. Saytzeff A (1885). Ueber die Oxydation der Oelsäure mit Kaliumpermanganat in alkalischer Lösung. J Prakt Chem 31:541-542; (1886). Untersuchungen aus dem chemischen Laboratorium von Prof. Alexander Saytzeff zu Kasan. 26. Ueber die Oxydation der Oel- und Elaïdinsäure mit Kaliumpermanganat in alkalischer Lösung. ibid 33:300-318.

Chapter 5
Into a New Century: Chemists Advancing the Legacies of Kazan', St. Petersburg, and Moscow

5.1 Introduction

By the end of the 1870s, Russian organic chemistry had reached the technical and intellectual stature needed to make it unnecessary for a student to leave Russia to receive advanced training in the discipline. There were vibrant schools of chemistry at St. Petersburg, Moscow, and Kazan', headed by Russian (and Russian-trained) Professors: Butlerov, Zinin, Borodin and Menshutkin at St. Petersburg, Markovnikov at Moscow, and Zaitsev at Kazan'. As noted in Chap. 1, the founding of new universities continued throughout the nineteenth: Warsaw University (1816), St. Vladimir University (Kiev, 1834), and the Imperial Novorossiisk University (Odessa, 1865) and the Tomsk Imperial (Siberian) University (1878) are important examples, and each these universities played a role in bolstering the development of organic chemistry in pre-revolutionary Russia.

Nevertheless, it was not until the twentieth century that the majority of students actually obtained all their chemical education in Russia: the experience of the *komandirovka* remained an attractive option for young chemists seeking to broaden their scientific and personal perspectives. With the coming of a new generation of Russian-trained chemists, coinciding with the second push to decentralize higher education in Russia, there also came something of a diaspora of chemical thought throughout the Russian empire. Students trained in Russia still sought study opportunities abroad, and many studied with eminent chemists in western Europe. With the passing of the older generation of chemists, individuals such as Victor Meyer (1848–1897), Wilhelm Ostwald (1853–1932), Adolph Strecker (1822–1871), and Johannes Wislicenus (1835–1902) became popular mentors for Russian students seeking advanced training. The trend, however, was changing, and more and more students received all their training within the confines of the Russian empire.

The major expansions of higher education in Russia were to the south, into the region now occupied by the Ukraine, and to the east—to Siberia. At the same time,

D. E. Lewis, *Early Russian Organic Chemists and Their Legacy*,
SpringerBriefs in History of Chemistry, DOI: 10.1007/978-3-642-28219-5_5,
© The Author(s) 2012

Warsaw University, which had been closed and reopened multiple times during the century, was subjected to intense Russification. In part, this was done as a way to try to stem the revolutionary activities of the Poles, who sought freedom for their homeland. It is worth remembering that Marie Curie's naming of the new element, polonium, after her homeland, was an act of rebellion, since the kingdom of Poland had ceased to exist everywhere except in the hearts and minds of the Poles.

Adolph Strecker (1822-1871)

Johannes Wislicenus (1835-1902)

Viktor Meyer (1848-1897)

Wilhelm Ostwald (1853-1932)

5.2 The Legacy of Kazan'

The organic chemistry school at Kazan' had been firmly established by Zinin and Butlerov before their departures for St. Petersburg, and their work had been consolidated by Butlerov's students, Markovnikov and—more importantly—Zaitsev, who assumed Butlerov's mantle at Kazan'. Another Butlerov student, Nikolai Nikiforovich Popov, became one of the inaugural professors at the Imperial University of Warsaw.

In a career spanning 4 decades, Zaitsev was responsible for the education of a number of important organic chemists; among these, three became eminent organic chemists in their own right, with named reactions or rules that have gone into organic chemistry textbooks: Yegor Yegorovich Vagner (Егор Егорович Вагнер, Georg Wagner, 1849–1903), Sergei Nikolaevich Reformatskii (or Reformatsky, Сергей Николаевич Реформатский, 1860–1934), and Aleksandr Yerminingel'dovich Arbuzov (Александр Ерминингельдович Арбузов, 1877–1968). In addition, Reformatskii's younger brother, Aleksandr Nikolaevich, became Professor of Analytical Chemistry in Moscow.

5.2.1 Kazan' University: The Twentieth Century

François Auguste Victor
Grignard (1871-1935)

Over the last third of the nineteenth century, Butlerov and Zaitsev had established Kazan' as a center of excellence in organic synthesis. Their organozinc syntheses had become the standard for the construction of alcohols, and this trend appeared to be certain of continuation well into the twentieth century. Then Victor Grignard published his synthesis with organomagnesium halides [1].

This one simple discovery ended the synthesis of most alcohols by means of organozinc nucleophiles for the next 8 decades [2], and dramatically altered the course of the Kazan' school of chemistry. The direction of the Kazan' school of chemistry was set for the next century by Zaitsev's student and successor, Aleksandr Yerminingel'dovich Arbuzov, who was the last student to publish the synthesis of an alcohol using the organozinc halide method of Zaitsev [3], and the first to publish a synthesis using organomagnesium, rather than organozinc nucleophiles [4].

5.2.1.1 Zaitsev's Successor: Aleksandr Yerminingel'dovich Arbuzov (1877–1968)

Arbuzov [5] was born to a lesser nobleman, Yerminingel'd Vladimirovich Arbuzov, and his wife, Nadezhda Aleksandrovna, a rural schoolteacher, in Arbuzov-Baran, a village in Kazan' province. He began his schooling in 1885, in the local, one-room school, but a year later his parents enrolled him in the Kazan' Classical Gymnasium. Arbuzov graduated from the Gymnasium in 1896, and entered Kazan' University as a student in the Physico-Mathematical Faculty. He graduated with the *Diplom* in May 1900, and in June the same year passed the test required for the degree of *kandidat* in natural sciences. He remained at Kazan' to pursue his research for the degree of M. Chem. As we have seen, it was clearly intended that he would pursue research along the lines of all Zaitsev's earlier students, but the higher yields and easier experimental procedure of the Grignard reaction meant that that project was now obsolete.

In response to these events, Arbuzov changed his focus to compounds of phosphorus, and in doing so, he set the course of chemical research at Kazan' for much of the rest of the century. In 1905, he submitted his dissertation for the degree of M. Chem. [6], entitled, "Concerning the structure of phosphorous acid and its derivatives." In this study, he prepared pure trialkyl phosphites, and discovered their catalytic isomerization to dialkyl alkylphosphonates [7].

Aleksandr Yerminingel'dovich
Arbuzov (1877-1968)
(Александр Ерминингельдович Арбузов)
photographed before 1915

In 1901, Arbuzov followed the advice of Zaitsev's colleague, Flavian Mikhailovich Flavitskii,[1] and he took a position as Assistant in the Department of Organic Chemistry and Agricultural Chemical Analysis in the Novo-Aleksandriya Institute of Agriculture and Forestry (Fig. 5.1), near Warsaw. In this, he followed the path of another Zaitsev student, Yegor Yegorovich Vagner. In 1906, Arbuzov was appointed to the Chair of the same department, and in 1907, he was awarded a *komandirovka* (rather like a modern sabbatical), which he spent in western Europe, working with Emil Fischer in Berlin, and Baeyer in Munich. He returned to Novo-Aleksandriya in 1910.

The institute occupied the Pulavski palace, on the right bank of the Vistula River, and like many Polish institutions, went through several cycles of opening, closing, and reopening. In 1862, it was transferred to the University of Warsaw, and in 1869 it reopened (along with the new Imperial University) as the Institute of Agriculture and Forestry. In 1893, it was reorganized yet again.

At the end of 1910, Zaitsev died, and his chair became vacant. Initially, two potential candidates for the position were identified: Vladimir Vasil'evich Chelintsev[2] (Владимир Васильевич Челинцев, 1877–1933), who chose to take a position at Moscow University instead, and Zaitsev's student, Aleksandr Nikolaevich Reformatskii[3] (Александр Николаевич Реформатский, 1864–1937), who also chose to remain at Moscow, and declined to compete for the position. Arbuzov's call to Kazan' was orchestrated to some degree by the same man who had recommended him to Novo-Aleksandriya: Flavian Flavitskii, who was now the senior Professor at Kazan'. Flavitskii's report on Arbuzov's work and potential

---

[1] Flavian Mikhailovich Falvitskii (Флавиан Михайлович Флавитский, 1848–1917) graduated from Khar'kov University, and moved to St. Petersburg where he worked under Butlerov until 1873. After taking his degree of *kandidat*, he moved to Kazan' as junior colleague to A. M. Zaitsev. He remained at Kazan' until his death, becoming Professor there in 1884. He was elected a corresponding member of the Academy of Sciences in 1907.

[2] Chelintsev was born in Saratov, and after finishing his secondary schooling here, he entered Moscow University, where he studied under Zelinskii. In 1900, he was awarded a *komandirovka*, which he spent with Barbier in Paris; here he met and worked with Grignard. From his return to Russia until 1918, he remained at Moscow University, becoming Professor of Chemistry in 1910, before returning to Saratov in 1918 as Professor at the new Saratov University (founded in 1909 as a result of an edict of Nicholas II). Here he spent the remainder of his career. His work was primarily in organomagnesium chemistry and heterocyclic chemistry. He was elected a Corresponding Member of the Academy of Sciences of the USSR in 1933.

[3] Like his older brother, Reformatskii studied under Zaitsev at Kazan' University, where he graduated in 1888. In 1889, he moved to the University of Moscow, where he delivered lectures in organic and inorganic chemistry, and on the history of chemistry. His M. Chem. research on rose oil was carried out under Markovnikov's direction. Following his graduation, he took a *komandirovka* to study with Viktor Meyer at Heidelberg. On his return to Russia, he began his independent work on the chemistry of aromatic aldehydes, and the reactions of allyl iodide and zinc on ethyl α-bromopropionate. Reformatskii was Professor of Chemistry at Shanyavskii People's University from 1901 to 1919, and at the second Moscow State University (originally the Women's Higher Courses) from 1918 onwards.

Fig. 5.1 The Novo-Aleksandriya institute of agriculture and forestry, in the Pulavski palace

was written in glowing terms. As a result, Arbuzov became Extraordinary Professor of Chemistry in September 1911.

Arbuzov was Professor of Chemistry at Kazan until 1930, and from 1922 to 1930 he served as Deputy Dean of the Physico-Mathematical Faculty. In 1930, he became Professor of Chemistry and Director of the Kazan' Technical Institute of the USSR Academy of Sciences; he became permanent Director of the Institute in 1945. In 1932, Arbuzov was elected a Corresponding member of the Academy, and in 1942, he was elected a Full Academician. In 1965, the Institute merged with the Kazan' Institute of Organic Chemistry, and the new body was named the A. E. Artbuzov Institute of Organic and Physical Chemistry in his honor. In 1948, he wrote a history of organic chemistry in Russia [8].

Beginning with his M. Chem. research in Poland, Arbuzov focused his attention on the chemistry of trivalent phosphorus. In 1914, he defended his dissertation for the degree of Dr. Chem. [9] at Kazan' University, and was promoted to Professor at the beginning of 1915. This dissertation reported his continued work in the area of rearrangements in phosphorus compounds. The lasting legacy of this work is the reaction that bears his name: the Arbuzov (or Michaelis-Arbuzov) rearrangement (Fig. 5.2).

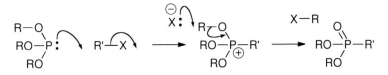

Fig. 5.2 The mechanism of the Arbuzov (Michaelis-Arbuzov) rearrangement

5.2.2 Warsaw University: The Kazan' Link

Warsaw University Main School had been closed in 1869, to be reopened in 1870 as the new Russian-language Imperial University of Warsaw. In order to begin operations, one of the first thing needed was a professoriate able to teach in Russian rather than Polish. Thus, they sought out chemists who had been trained in Russia, and students of Butlerov were obviously at a premium for the positions. Over the course of the next decade, two Kazan' students, both of whom had studied under Butlerov, became Professors at the University. The first of these was Aleksandr Nikiforovich Popov (ca. 1840–1881) [10], who accepted an invitation in 1869 to take a position as Professor of Organic Chemistry at the university, and the second was Yegor Yegorovich Vagner (Егор Егорович Вагнер, Georg Wagner, 1849–1903). It was Popov's departure from Kazan' that cleared the way for Za-itsev to become Markovnikov's junior colleague there.

5.2.2.1 Aleksandr Nikiforovich Popov (ca. 1840–1881)

Popov, who is pictured in Fig. 5.3 along with Butlerov and several Butlerov students, was born in Vitebsk province, and studied at Kazan' University from 1861 to 1865, where he came under the influence of Butlerov at a time when the latter was teaching his version of structural theory, and working to establish it as the standard version of the theory. Popov presented a dissertation for the degree of *kandidat* in cameral science in 1865 [11], and he continued at Kazan as an assistant in Butlerov's laboratory until he submitted his dissertation for the degree of Dr. Chem., in 1869 [12]. He continued at Kazan' until his call to Warsaw in 1869.

Like his immediate supervisor, Markovnikov, Popov had a superb grasp of the nuances of structural theory and its use, but he combined theoretical brilliance with excellent practical skills. Unfortunately for Warsaw University, Popov's health was never robust, and he died quite young without realizing his full potential. Some idea of that potential may be gauged from his *kandidat* dissertation on structural theory, in which he provided experimental evidence of the equivalency of the four bonds to carbon. His first paper [13], which was published while he was still a student with Butlerov, was the first of a series [14] describing the oxidation of ketones with chromic acid that eventually formed the basis for his Dr. Chem dissertation. In this paper, he provides unequivocal evidence (Fig. 5.4) that the synthesis of a ketone from the two different possible combinations of an acid

Fig. 5.3 Butlerov and a group of his students, 1867–1868. *Rear row*: unidentified; A. N. Popov; A. M. Zaitsev; G. N. Glinskii. *Front row*: unidentified; unidentified; A. M. Butlerov; A. I. Loman; unidentified; unidentified

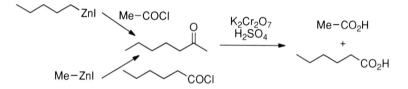

Fig. 5.4 Popov's demonstration of the equivalency of the bonds to the carbonyl carbon

chloride and an organozinc reagent gives the same compound, contradicting Kolbe's view that the two bonds to the carbonyl carbon would be distinguishable by their method of construction.

Popov's major contributions were in the use of destructive oxidation by chromic acid ($K_2Cr_2O_7$–H_2SO_4) for the determination of structures of alcohols, ketones, and alkylated arenes. During the winter of 1871–1872, Popov studied in Kekulé's laboratory in Bonn. Here he worked with Theodor Zincke (1843–1928), and the two chemists were able to show that the oxidation of alkylbenzenes occurred at the benzyl carbon to give radicals that were then oxidized further [15].

Popov studied the oxidation of organic compounds with chromic acid, and deduced rules for determining the structures of alcohols, ketones, acids, and hydrocarbons based on the products of these reactions. One of Popov's rules, which was formalized in his thesis for the degree of Dr. Chem. [3], states that the oxidation of an unsymmetrical diketone, RCOR', by chromic acid, leads to the carboxylic acid, RCO_2H, where the probability of the alkyl group attached to the carbonyl group decreases in the order Ph $\approx$ R_3C > Me > RCH_2 > R_2CH > $PhCH_2$. It is not accidental that this order of reactivity corresponds to the stability of the isomeric enol tautomers of the ketones, although Popov could not know this at the time.

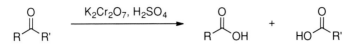

Popov's work foreshadowed Zaitsev's Rule for elimination, but his contribution was not recognized outside the Russian Empire (just as Zaistev's rule went unattributed until the 1960s). In fact, Popov's rule was forgotten in favor of Zaitsev's rule during the Soviet era. It is also worth mentioning that, although he certainly used Popov's results in his formulation of his rule for elimination, for some reason Zaitsev did not acknowledge Popov's prior work [16].

5.2.2.2 Yegor Yegorovich Vagner (Georg Wagner, 1849–1903)

Yegor Yegorovich Vagner [17] was one of the most perceptive organic chemists of his generation. He is much better known in the west under the German form of his name, Wagner, despite the fact that both he and his father deliberately used the name *Yegor* (a Russian version of the name Georgii) to emphasize that they were Russian, despite their German origins. In this chapter, we will use the transliteration of the Russian, rather than the German form of Vagner's name, since that was the stand that he, himself, took.

Vagner was the grandson of August Wagner, a Prussian pharmacist who had migrated to Kazan' while still a young man. His son, Yegor Avgustovich, was enrolled in the Juridicial-Economics Faculty of Kazan' University and after graduating became a government official in the Samarra province before he returned to Kazan', where his son, Yegor Yegorovich was born in 1849. Vagner's mother, Aleksandra Mikhailovna Lvov, was the daughter of the director of the Kazan' gymnasium, and came from an old, noble family. When Yegor was just a year old, she died of tuberculosis, so his grandmother assumed many of the duties of a mother.

Once he reached school age, young Yegor was sent to boarding school in Livonia, on the Baltic coast. However, at 16, he had had enough of this school, and he ran away—by train to Nizhni Novgorod, then by baggage train to Kazan'. His indulgent father is said to have greeted him with, "Well, brother, you're a perfect Lomonosov, except the opposite: he fled by baggage train *to* his studies, and you *from* learning!" Vagner finished his schooling at home, and passed the examinations required for entry into the university.

Yegor Yegorovich Vagner
(Егор Егорович Вагнер, Georg Wagner,
1849-1903) in his laboratory

Following his father's advice, Yegor Yegorovich Vagner entered Kazan' university in 1867 as a student in law. As a cameral student, he was required to take
2 years of chemistry, and he attended lectures by both Markovnikov and Zaitsev.
This was enough to cause him to change his career, so that after 2 years of study,
he transferred from the Juridicial-Economics Faculty to the Physico-Mathematical
Faculty and started over. He began studying under Zaitsev, and during this time he
began his work with organozinc nuceophiles. His first publication with Zaitsev was
a synthesis of secondary alcohols by the reaction between ethyl iodide, zinc metal,
and ethyl formate to give diethylcarbinol [18]. Later, while at St. Petersburg,Vagner would expand this method to the synthesis of unsymmetrical secondary alcohols by the reaction between an alkylzinc iodide or a dialkylzinc and
an aldehyde, a method he perfected in Warsaw [19].

In 1875, Vagner was awarded a 2 year *komandirovka* to prepare for the
professoriate. However, instead of traveling to western Europe to study, Wagner
chose to continue studying with Zaitsev for the first year, and then to study with
Butlerov at St. Petersburg. This decision by Vagner marks the maturation of
organic chemical education in Russia because, instead of needing to go abroad to
obtain his capstone educational experience, he could remain in Russia to study
with chemists with international reputations. It is also a testament to Zaitsev's
generosity towards his students: he did not stand in the way of his most brilliant
and productive student as he tried to better himself by moving to a more

prestigious institution. In St. Petersburg, Vagner worked under Butlerov, and even more closely with Menshutkin. In 1876, he became Menshutkin's assistant in the analytical chemistry laboratory, where he remained until 1882.

Kazan'   St. Petersburg and Warsaw

In 1882, Wagner moved to the Novo-Aleksandriya Institute of Agriculture and Forestry as Professor of Chemistry. At the Institute, Wagner worked on his research for the degree of M. Chem.; he submitted his dissertation to St. Petersburg in 1886 [20].

While at the Institute of Agriculture and Forestry, Vagner's studies of alcohols were extended to their oxidation by chromic acid. This also led Vagner to begin a systematic study of the chromic acid oxidation of ketones [21], extending the work begun by Popov. What he found was that, in contrast to Popov's rule, which stated that the oxidation gave two acids by breaking the bond between the carbonyl group and the smaller of the two groups, oxidation of dialkyl ketones could give four different carboxylic acids, with the proportions of each depending on the size of the alkyl group and the number of hydrogen atoms at what we now term the α carbon.

On graduating with his M. Chem. degree, he moved to the Imperial University of Warsaw as Extraordinary Professor. Here he began working on the oxidation of unsaturated compounds with aqueous potassium permanganate. The oxidation of alkenes with permanganate had been studied by Zaitsev, who studied the oxidation of oleic and elaidic acids [22], Tanatar,[4] who studied the oxidation of fumaric and maleic acids [23], and other contemporary organic chemists [24].

Vagner studied the reaction in depth [25], and established the critical experimental parameters: that it was necessary to keep the concentration of the permanganate solution below 4%, and to keep the pH high to avoid over-oxidation of the diol product. He also established that the double bond is not cleaved by permanganate, and that the reagent is specific for alkenes. Since benzene does not react with this reagent, this also made Vagner question Kekulé's structure for benzene. Vagner's papers described the work that he submitted to Warsaw University in his successful dissertation for the degree of Dr. Chem. [26].

---

[4] Sevastian Moiseevich Tanatar (Севастиан Мойсеевич танатар, 1849–1917) was a student of Aleksandr Andreevich Verigo (Александр Андреевич Вериго, 1837–1905), who, along with Sokolov, had organized the chemistry laboratories at Odessa. During a period abroad after his graduation from Odessa University, he became interested in the then fashionable topic of the origin of the isomerism of fumaric and maleic acids. He reported the oxidation of these acids with potassium permanganate to give dihydroxyacids that were later shown to be *meso* and *dl* tartaric acids. Tanatar became Professor at Odessa in 1896. His later research was with peroxides.

Fig. 5.5 The Wagner
(Vagner)–Meerwein
rearrangement

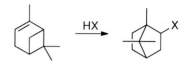

During the final stage of Vagner's career, he turned his attention to the problem of the structure of the terpene hydrocarbons. This was one of the hot fields of organic chemistry in the late nineteenth century; many, including Nobel Prize-winners, Otto Wallach[5] and Adolph von Baeyer,[6] focused their attention here. Vagner's contribution, which ultimately led to the correct structure of several terpene hydrocarbons, was to apply the permanganate oxidation to the problem of locating the double bonds in the hydrocarbons. Using this reaction, he was able to deduce correct structures for a number of terpenes, including limonene [27], terpineol [28], α- and β-pinene [28, 29], camphene [30], and bornylene [31]. During his structural studies of terpenes, Vagner deduced the relationship between the pinane and bornane carbon skeletons by studying the reaction of α-pinene with hydrogen chloride and hydrogen iodide to give the corresponding bornyl or iso-bornyl halide, and suggested the rearrangement that relates them [21, 32] (Fig. 5.5). The involvement of carbocations in the mechanism was proposed by Meerwein[7] 2 decades later [33].

Vagner did nothing by halves. Today, he would no doubt be characterized as a workaholic: during the time he was doing the research for his M. Chem. dissertation, he was typically in his laboratory before anyone else, and he left after everyone else, and the light in his laboratory often burned late into the night. However, Vagner was not all business: Kazan' students in the 1870s had a passion for drama, and Vagner was no exception. He was not only a regular visitor to the

---

[5] Otto Wallach (1847–1931) received the Nobel Prize for chemistry in 1910 for his research on terpenes. In the quarter century between 1884 and 1909, he published over 100 papers on the chemistry of terpenes. In 1887, he proposed—albeit heistantly—what became known as the "isoprene rule" [Wallach (1887). Zur Kenntniss der Terpene und der ätherischen Oele; Fünfte Abhandlung. Justus Liebigs Ann Chem 239:1–54].

[6] Johann Friedrich Wilhelm Adolf von Baeyer (1835–1917) received the Nobel Prize in 1905 for his work in synthetic organic chemistry, especially the synthesis of organic dyes and heterocyclic compounds.

[7] Hans Leberecht Meerwein (1879–1965) took his Ph.D. from Bonn in 1903. After three semesters as a laboratory assistant at the technical university at Charlottenburg, he returned to Bonn as an assistant in inorganic and analytical chemistry; in 1914 he was promoted to a faculty position. During World War I, Meerwein commanded an airship company. In 1923 he moved to the University of Königsberg as professor of Chemistry, and in 1928 he was simultaneously offered Professorships at the Universities of Leipzig and Marburg; he chose Marburg, where he helped to plan the Chemical Institute, becoming its first director. He remained there until his death. Meerwein's major contributions to modern organic chemistry include the mechanism of the 1,2–rearrangements of cations), the selective reduction of aldehydes and ketones by metal alkoxides (especially aluminum alkoxides) now known as the Meerwein-Pondorff-Verley reduction, and the synthesis of trialkyloxonium salts (called Meerwein salts).

city theater, but he himself took an active part in amateur theatricals: there are extant, posters for shows in which he appeared as a performer a wide variety of roles. This love of theater and fun was evident throughout his life: the chemist Nikolai Ipatieff tells of a night during a conference in Kiev, when Vagner took a group of chemists from the conference to a cabaret on Trukhanov Island in the Dnieper River. To quote Ipatieff [34]:

> Just before I left I spent most of the day and night with a group of chemists headed by Wagner himself. We dined at the Hotel Continental, the meal being supplemented by considerable drinking of our national vodka and by a most lively discussion on the papers presented at the meeting. About four o'clock in the afternoon our chemical leaders, being in a very happy frame of mind, decided to go to Trukhanov Island on the Dnieper River, where there was a restaurant and a cabaret. At the island we continued our discussion, drinking tea and cognac, which proved as popular as the vodka. After tea we saw the cabaret performance probably with more vigor than politeness. Then the celebration continued in a separate room of the restaurant with me, the only one still sober, in charge. Wagner was the life of the party and I must note that few can achieve the happy and pleasant state of mind that he could. We returned to Kiev at four o'clock in the morning…

In 1875, Vagner married Vera Alexandrovna Barkhatova (1855–1880); her death 5 years later left him with two young sons, Egor and Dmitrii, to raise. He remarried his son's governess, Alexandra Afanas'evna Afanas'eva; they had a son, Roman. Throughout his life, Vagner struggled with obesity, and the attendant heart disease, but his death was due to post-operative complications of surgery for what today we would call colorectal cancer, less than 2 weeks before his 54th birthday. Vagner was a brilliant chemist who died too young: von Baeyer conceded that he had been the first to establish the correct formulas for the menthane and pinene classes of terpenes by correctly interpreting the results of others [35], and Meerwein commented on the "wondrous sharpwittedness" of Vagner [36].

5.2.2.3 Nikolai Aleksandrovich Prilezhaev

One of the last of Vagner's students at Warsaw was Nikolai Aleksandrovich Prilezhaev (Николай Александрович Прилежаев, 1872–1944) [37]. The son of a priest, and the nephew of St. Petersburg chemist Aleksei Yevgrafovich Favorskii, whom we will discuss later in this chapter, Prilezhaev graduated from the seminary in 1895. Although destined for life as a priest, his father's death allowed him to pursue a career as a scientist instead, and in 1896 he entered the Natural Sciences Department of the Physics–Mathematics Faculty of Warsaw University. He graduated in from this department in 1900. Following his graduation, he received an appointment as a junior assistant in the Chair of Organic Chemistry at the Warsaw Polytechnic Institute, then occupied by Vagner. His choice to pursue organic chemistry was very likely influenced by his uncle. Prilezhaev's work with Vagner concerned the oxidation of alkenes, and this concentration on oxidation continued after Vagner's death.

Fig. 5.6 The Prilezhaev
(Prileschajew) reaction

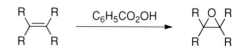

Nikolai Aleksandrovich Prilezhaev
(Николай Александрович Прилежаев,
1872-1944)

In 1909, Prilezhaev published the first account of his discovery of the epoxidation of alkenes by perbenzoic acid [38] (Fig. 5.6). This discovery was the central theme of his M. Chem. dissertation, "Organic peroxides and their application to the oxidation of unsaturated compounds" [39], defended at Warsaw in 1912. On graduating M. Chem., he was appointed Professor of Organic Chemistry at Warsaw University. Three years later, he moved to Kiev Polytechnic Institute as Professor, and finally, in 1924, he moved to Belarus University, becoming one of the organizers of the chemistry department there. In 1933, he was elected a Corresponding Member of the USSR Academy of Sciences, and in 1940 he was elected a full member of the Belarus Academy of Sciences in 1940.

5.2.3 The Kiev Connection with Kazan'

Kiev, the capital and largest city of modern Ukraine, is one of the oldest cities in eastern Europe. In 1667, it was incorporated into the Russian Empire, but it retained a degree of autonomy. An important religious center for Christian pilgrims since the Christianization of the Kievan Rus by the ruler, Vladimir the Great, late in the tenth century, Kiev was of marginal economic importance until the nineteenth century, when Russification of the region was accompanied by migration of Russians into the area. The founding of St. Vladimir University (now Taras Shevchenko National University of Kyiv) followed the influx of Russians, rather than preceding it. The rise of the chemistry program there to prominence can be traced to yet another graduate of the Kazan' school: Sergei Nikolaevich Reformatskii.

5.2.3.1 Sergei Nikolaevich Reformatskii (1860–1934)

Sergei Nikolaevich Reformatskii
(Сергей Николаевич Реформатский,
1860-1934)

Sergei Nikolaevich Reformatskii [40] was the older of two brothers who both studied chemistry under Zaitsev at Kazan'. The two brothers followed remarkably similar career paths. Both were the sons of a priest, and after graduating from the Kostroma Spiritual Academy, entered Kazan' University as a students in the Physico-Mathematical Faculty, where they each entered Zaitsev's laboratory.

In 1882, Reformatskii graduated with the degree of *kandidat* and the gold medal for his dissertation on the hydrocarbon $C_{10}H_{18}$ produced by dehydration of allyl dipropyl carbinol (4-propylhept-1-en-4-ol) [41]. On his graduation, he was appointed caretaker of the laboratory museum. Reformatskii continued working at Kazan', carrying out research into the synthesis of glycols by means of hypochlorous acid oxidation of alkenes. He published one paper in this area [42] before

Fig. 5.7 The Reformatskii reaction

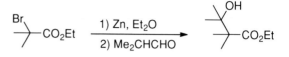

submitting his dissertation for the degree of M. Chem. [43] in 1889. Upon his graduation, he was awarded a *komandirovka*. He spent his years abroad in the laboratories of Viktor Meyer in Göttingen and Heidelberg, and with Wilhelm Ostwald at Leipzig. During this time, he worked on extending the scope of the organozinc chemistry developed by Butlerov and Zaitsev. This work formed the cornerstone of his doctoral dissertation, and led to his lasting fame. In 1890, on his return from abroad, he defended his dissertation, "The action of a mixture of zinc and monochloroacetic ester on ketones and aldehydes" [44] at Warsaw University. In 1891, Reformatskii was appointed to the Chair of Organic Chemistry at St. Vladimir University, in Kiev, a post he occupied until his death. In 1928, Reformatskii was elected a Corresponding member of the Academy of Sciences of the USSR.

Unlike the other reactions of zinc alkyls with carbonyl compounds, which were supplanted by the Grignard reaction, the Reformatskii reaction has stood the test of time [45]. The first paper [46] in this series was followed by a stead outflow of work describing the synthesis of β-hydroxyacid derivatives [47]. The reaction remains a mainstay of synthesis because of the tolerance of the Reformatskii reagent to a wide array of functional groups, and because the reagent reacts with ketones to give the β-hydroxyester (unlike the aldol additions of lithium enolates to ketones, where the equilibrium is frequently unfavorable), as shown by the reaction between acetone and the Reformatskii reagent from α-bromoisobutyric acid [48] (Fig. 5.7):

As implied above, Reformatskii's independent research retained the essential character of the chemistry begun at the Kazan' school by Butlerov and Zaitsev. The facile formation and reactions of allylzinc iodides with ketones made examining the reactions of the structurally analogous α-halocarbonyl compounds an obvious next step, and it was to Reformatskii's benefit that this led to a highly useful reaction. Of all Zaitsev's students to attain prominence, only Reformatskii remained close to his professional roots throughout his career.

5.3 The Legacy of Moscow

5.3.1 The Markovnikov School

Organic chemistry at Moscow became much more prominent with the arrival of Markovnikov as Professor of Organic Chemistry in 1873. Not only did he work hard to improve the facilities there, but he also attracted an energetic group of students to his banner, just as he had done at Kazan' and Odessa. At least three of his students achieved international stature during their careers, and a fourth, Mikhail Ivanovich Konovalov,[8] was instrumental in helping one of the three to complete his education.

Markovnikov's ouster in 1893 did not end his influence. His later work on the Caucasus oils inspired many of his students to pursue hydrocarbon chemistry in their own careers. Markovnikov's students became influential members of the chemical community in Moscow: at the Petrovskaya Academy of Agriculture and Forestry (now the Timiryazev Agricultural Academy), which was founded in 1865, and at the Women's Higher Courses (which eventually became the Second Moscow State University), which had been founded in 1872, when historian Vladimir Ivanovich Guerrier (Владимир Иванович Герье, 1837–1919 was granted permission to open an institution of higher for women. His student, Kizhner, became Professor of Chemistry at Tomsk, the first university in Siberia. His students also taught at the short-lived Shanyavskii People's University. This institution was unusual because it was not an official university, and could not grant degrees. It was founded in 1911 after over a 100 professors resigned from Moscow University in protest over the actions of the Minister of Education, Lev Aristidovich Kasso,[9] dismissing three professors (Manuilov, Menzbir, and Minakov) for anti-tsarist activity. Many of these same professors then set up their own university in competition with Moscow University. Despite its unofficial status, Shanyavskii People's University was a very influential institution.

5.3.1.1 Nikolai Yakovlevich Dem'yanov (1861–1938)

Nikolai Dem'yanov (Нколай Яковлевич Демьянов, also spelled Demjanoff, Demjanow, and Demjanov) was born in Tver, northwest of Moscow, and educated at the fourth Classical Gymnasium in Moscow. In 1882, he entered Moscow University, and in 1886 he completed his *kandidat* dissertation, "On dextrin," under Markovnikov's direction [49].

---

[8] Konovalov (Михаил Иванович Коновалов, 1858–1906) was the son of a farmer who left his home at age 15 to enter the Gymnasium in Yaroslavl. He graduated from the Gymnasium in 1880, and entered Moscow University as a student in chemistry. He entered Markovnikov's laboratory, and in 1884 he graduated, and, Markovnikov recommended that he be retained by the university. In 1889, he submitted his M. Chem. dissertation, "Naphthenes, hexahydrobenzenes and their derivatives."

[9] Lev Aristidovich Kasso (Лев Аристидович Кассо, 1865–1914) was a wealthy landowner who was educated abroad as a lawyer. He held positions at Dorpat (1892), Khar'kov (1895), and Moscow (11899) before becoming Director of the Imperial Lyceum (1908). In 1910, he was appointed to the Ministry of Education, and in 1911, he became minister. His ruthless reactionary policies included banning student unions and assemblies, and dismissing progressive professors. He intensified the surveillance of all in higher education, and, in 1912, he expelled all women students from the Higher Medical Courses in St. Petersburg.

On graduating, Dem'yanov moved to the Petrovskaya Agricultural Academy in Moscow as Assistant to Gustavson.[10] He taught organic and biological chemistry, and pursued research towards his M. Chem. and Dr. Chem. degrees. Gustavson's interests were in the reactions of electrophiles with hydrocarbons [50], and in the chemistry of cyclopropane derivatives [51], and so it was natural that Dem'yanov should also follow his lead. His first papers were published jointly with Gustavson, and concerned allene and small-ring compounds [52].

Nikolai Yakovlevich Dem'yanov
(Николай Яковлевич Демьянов, 1861-1938).
Also: Demyanov, Demjanoff, Demjanow, Demjanov

Dem'yanov spent his entire career at the Agricultural Academy. In 1893, he was promoted to Adjunct Professor of Chemistry, and in 1898 he was promoted to Professor of Chemistry. His dissertation for the master's degree [53], which was successfully defended at St. Petersburg University in 1895, was his first report of the reaction of nitrous acid with amino compounds. In 1899, he defended his dissertation for the degree of Dr. Chem. [54] at Moscow University. This dissertation extended his studies of the reactions of nitrogen oxyacids and oxides with organic compounds to reactions of nitrogen dioxide, especially, with alkenes.

Dem'yanov's lasting fame came from his studies of the reaction of cyclobutylmethylamine with nitrous acid, a reaction that gives a mixture alcohol products,

---

[10] Gavril Gavrilovich Gustavson (Гаврил Гаврилович Густавсон, 1843–1908) was born and educated in St. Petersburg, where he studied under Mendeleev and Butlerov. In 1875 he was appointed Professor of Organic and Agronomic Chemistry at the Petrovskaya Academy of Agriculture and Forestry, in Moscow; in 1892 he returned to St. Petersburg, where he taught in the Higher Courses for Women (later the Second Moscow State University).

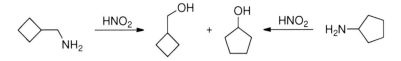

Fig. 5.8   The Demjanov rearrangement

including cyclopentanol [55]. This ring expansion, and the subsequent observations by Dem'yanov that one could use the same reaction to contract the ring [56], was the first example of a carbocation reaction that involves what today is termed a non-classical carbocation [57] (Fig. 5.8).

In 1937, a very similar rearrangement of β-aminoaclohols to give carbonyl compounds was reported by French chemist, Marc Tiffeneau[11] [58]. The advantage of the Tiffeneau rearrangement over the Demyanov rearrangement is that it allows medium-sized rings to be prepared by ring expansion (the formation of the carbonyl group provides much of the driving force). The two reactions are now generally combined, and known as the Tiffeneau-Demjanov rearrangement.

5.3.1.2 Nikolai Matveevich Kizhner

Nikolai Matveevich Kizhner (Николай Матвеевич Кижнер, or Kishner, 1867–1935) was born and educated in Moscow, where he graduated from the Gymnasium, and entered Moscow University in 1886. From 1890 to 1893, Kizhner worked in Markovnikov's laboratory, and from 1893 to 1898 he taught quantitative analysis and special courses in organic chemistry. In 1895, he defended his dissertation for the degree of M. Chem. [59] at St. Petersburg. Five years later, Kizhner defended his Dr. Chem. dissertation [60], in which he established the structure of hexahydrobenzene as cyclohexane instead of methylcyclopentane, at Moscow. In 1901, Kizhner took the Chair of Organic Chemistry at the Tomsk Technological Institute, where he quickly set up a well-equipped laboratory. However, his productivity was cut short by a serious illness (described in one account as "gangrene of the limbs") that, several operations later, left him disabled. He remained at Tomsk for 13 years. In 1906, on the orders of the Governor-General of Western Siberia, he was exiled for a year because of his progressive views, before being permitted to return to the Institute. In 1914, again against his will, but to improve his health, he returned to Moscow as Professor at the Shanyavskii People's University. In 1918, he

---

[11] Marc Émile Pierre Adolphe Tiffeneau (1873–1945) earned his Dr. en Sc. in 1907, and his Dr. en Méd. in 1910, and became Agrégé in the Faculty of Medicine at Paris the same year. His first teaching assignment was in pharmacology in the Faculty of Medicine; it was not until a decade and a half later (1924–1926) that he taught a course in chemistry in the Faculty of Science. In 1927, he was named Professor of Pharmacodynamics in the Faculty of Medicine. Tiffeneau received the Prix Jecker in 1911 and 1923, and was named Chevalier (1923) and Officier (1938) of the Légion d'honneur.

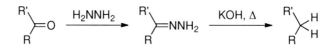

Fig. 5.9 The Wolff–Kishner reduction

became Director of the Aniline Trust Research Institute; in 1928 he was elected a Corresponding Member of the Academy of Sciences of the USSR, and in 1934 he was elected a Full Academician.

Nikolai Matveevich Kizhner
(Николай Матвеевич Кижнер, 1867-1935)

Kizhner's scientific legacy is based on the chemistry of hydrazine and similar derivatives of ammonia. In 1911, he published the method for the deoxygenation of ketones by heating their hydrazones with potassium hydroxide [61]. A similar reaction reported by Ludwig Wolff [62] a year later. The Wolff–Kishner reduction (Fig. 5.9) has been a mainstay of organic synthesis for a century now, with several variants of the reaction—most notably the Huang Minlon modification [63]—being developed over the years.

5.3.1.3 Aleksei Yevgen'evich Chichibabin

If ever a career could be used to model how perseverance can overcome adversity, the career of Aleksei Yevgen'evich Chichibabin (Алексей Евгеньевич Чичибабин, 1871–1945) must surely be a candidate. From difficult economic circumstances, he rose to prominence in Soviet organic chemistry with no less than three reactions and a hydrocarbon bearing his name [64].

Aleksei Yevgen'evich Chichibabin
(Алексей Евгеньевич Чичибабин,
1875-1941)

Chichibabin was born in Kuzemino, near Poltava in the Ukraine. At age 17, he moved to Moscow to begin his studies at the university, living at a boarding house for penurious students, and working to support himself. Despite these difficulties, he received his *diplom* in 1892. His path forward was anything but straightforward, however.

His chosen mentor, Markovnikov, had fallen victim to university politics, and Markovnikov's successor, Zelinskii, wanted nothing to do with the student of his predecessor; this meant that Chichibabin's student days at Moscow University were over. Instead, he moved to the Agricultural Institute, where he studied under Konovalov, himself a former Markovnikov student, who proposed a project on the nitrating activity of dilute nitric acid on the higher homologues of pyridine. During this time, Chichibabin supported himself as a journalist. Konovalov's departure for Kiev in 1899 once again left Chichibabin without a research mentor, but 2 years later, he managed to obtain a research position as an Assistant in the laboratory of Ivan Alekseevich Kablukov,[12] yet another student of Markovnikov and Butlerov. In 1903, Chichibabin submitted his M. Chem. dissertation [65], which was approved despite Zelinskii's negative opinion of the work (based mainly on the fact that Chichibabin was self-directed instead of belonging to an established research group). Chichibabin's doctoral dissertation [66] concerned a question that he had become interested in during 1904: the nature of free radicals.

In 1905, he was chosen Professor of Inorganic Chemistry at Warsaw, but the working conditions and atmosphere there soon determined him to return to his position in Moscow. In 1908, he became Professor of Chemistry and Director of the Moscow Higher Technical School; from 1909–1929 he served intermittently as Dean of the Faculty. His contributions to the development of the Soviet chemical industry were significant: he was instrumental in organizing the Russian pharmaceutical industry during World War 1, establishing the first Russian factory of pharmaceutical products at the Technical School. In 1929, he was elected a member of the Academy of Sciences of the USSR.

In 1930, Chichibabin's only daughter was killed in an oleum explosion, and the fact that it could have been prevented determined him to leave Russia. With his wife, he moved to Paris, where he worked with Tiffeneau at the Hôtel Dieu, and later at the Collège de France. At the same time, he continued his association with industry, becoming a Director of the Scientific Department of the Établissements Kuhlmann, a major French manufacturer of dyes and fine chemicals, and a scientific advisor to the Schering and Roosevelt Company in New York. As a result of his refusal to return to Russia, he was stripped of his Soviet citizenship and his membership in the Academy of Sciences in 1936. His status as an Academician was restored, posthumously, by the Academy in 1990.

Chichibabin's first contribution to organic chemistry was the development of a general synthesis of aldehydes (Fig. 5.10) by the reaction between a Grignard reagent and triethyl orthorformate [67], but this was rapidly eclipsed by his contributions to the chemistry of pyridine and its homologues. In 1906, he reported a general synthesis of pyridines (Fig. 5.11) by means of a multicomponent condensation reaction [68], now known as the Chichibabin pyridine synthesis, in which three molecules of an enolizable aldehyde condense with one molecule of

---

[12] Kablukov (Иван Алексеевич Каблуков, 1857–1942) was a Russian physical chemist who studied under Markovnikov (*kandidat* 1880, Moscow) and Butlerov (M. Chem. 1881, St. Petersburg). He became Docent at Moscow in 1885. Despite his early exposure to organic chemistry, a *komandirovka* with Ostwald in 1889 set him on the path to a physical chemistry career. He defended his Dr. Chem. dissertation on the theories of van't Hoff and Arrhenius in 1891, and in 1903 was elected Professor of the Agricultural Institute. Kablukov's work was in thermodynamics: he demonstrated that the heats of formation of isomeric organic compounds were not identical, and, independently of Kistiakowsky, he introduced the concept of ion solvation. He was elected to the Academy of Sciences of the USSR in 1932, and received both the Order of Lenin and the Order of the Red Banner of Labor.

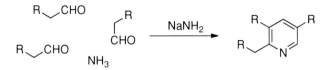

Fig. 5.10   The Chichibabin aldehyde synthesis

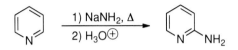

Fig. 5.11   The Chichibabin pyridine synthesis

Fig. 5.12   The Chichibabin reaction

ammonia to give a 2,3,5-trisubstituted pyridine. In 1914 he reported the reaction (Fig. 5.12) between pyridine and sodium amide to give 2-aminopyridine [69], a reaction known as the Chichibabin reaction.

5.3.2 Markovnikov's Successor: Nikolai Zelinskii and his Students

Nikolai Dmitrievich Zelinskii
(Николай Дмитиевич Зелинский,
1861-1953)

Fig. 5.13 The Hell–
Volhard–Zelinskii reaction

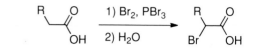

Nikolai Dmitrievich Zelinskii (Николай Дмитриевич Зелинский, 1861–1953) became a giant of Soviet chemistry. Zelinskii was born in Tiraspol, in modern Moldova, and educated at the Imperial Novorossiisk University in Odessa. He graduated with the degree of *kandidat* in 1864, and spent the next 2 years studying in western Europe, with Wislicenus at Leipzig and Viktor Meyer in Göttingen. On his return to Russia in 1887, he became Privatdocent at Odessa. In 1889, he defended his M. Chem. dissertation, "On the question of isomerism in the thiophene series" [70], and 2 years later he defended his Dr. Chem. dissertation, "Investigations of the phenomenon of stereoisomerism in saturated carbon compounds " [71]. In 1893, after Markovnikov's ouster, Zelinskii was appointed Extraordinary Professor at Moscow University. The relationship between the two chemists was never good, in part because Zelinskii had taken over Markovnikov's investigations in the laboratory without seeking Markovnikov's permission, which Markovnikov viewed as a violation of scientific ethics.

Zelinskii spent the remainder of his career at Moscow, except for the period 1911–1917, which he spent as Director of the Central Laboratory of the Ministry of Finance and as Professor in a sub-department of the St. Petersburg Technological Institute, after resigning as part of the mass protest against the actions of Minister of Education, Kasso, subverting overturning the university's autonomy. He was elected a Corresponding member of the Academy of Sciences in 1926, and a Full Academician in 1929. In 1934, he organized the Organic Chemistry Institute of the Academy of Sciences in Moscow, and in 1953 this institute was named in his honor.

Zelinksii's work was prolific and varied—at the time of his death, he had authored over 500 scientific papers. In 1887, he published the account of the bromination of carboxylic acids (Fig. 5.13) that now bears his name, along with those of Carl Magnus von Hell (1845–1926) and Jacob Volhard (1834–1910) [72]. In 1908, he published a synthesis of α-aminoacids with his student, Stadnikov, that bears both their names [73]. In this paper, he demonstrated that α-aminonitriles could be formed from aldehydes by a useful modification of the Strecker synthesis (which had involved using ammonia and hydrogen cyanide) by using aqueous potassium cyanide and ammonium chloride (Fig. 5.14).

In addition to his work with thiophenes, and the work that led to the two reactions that bear his name, Zelinskii also carried out wide-ranging research in a variety of industrial or applied organic chemistry topics. His work on adsorption led to the development of gas masks used by the allied forces during World War 1, and his activities in the study of catalysis were widespread, including the first smooth catalytic dehydrogenation of cyclohexane to benzene [74], which presaged modern catalytic reforming of petroleum, and the condensation of acetylene to benzene in the presence of activated charcoal [75].

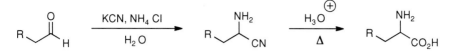

Fig. 5.14 The Zelinskii–Stadnikoff modification of the strecker amino acid synthesis

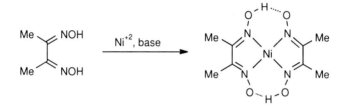

Fig. 5.15 The Chugaev xanthate pyrolysis (Chugaev elimination)

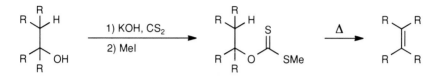

Fig. 5.16 Complexation of nickel (II) by dimethylglyoxime

5.3.2.1 Lev Aleksandrovich Chugaev (1873–1922)

Lev Aleksandrovich Chugaev (Лев Александрович Чугаев) was the younger half-brother of another eminent Russian chemist of the period encompassing the turn of the twentieth century: Vladimir Nikolaevich Ipatieff (Владимир Николаевич Ипатьев, 1867–1952). Chugaev was born in Moscow, and educated at Moscow University, where he graduated with the degree of *kandidat* in 1895, under Zelinskii. He received his M. Chem, in 1903 for a dissertation, "Investigations in the field of terpenes and camphors" [76], during which time he discovered and developed his xanthate pyrolysis method (Fig. 5.15) for the dehydration of alcohols [77]. He received his Dr. Chem. 3 years later for a dissertation, "Investigations in the field of complex compounds" [78], during which he discovered dimethylglyoxime as a highly specific reagent for the gravimetric determination of nickel [79] (Fig. 5.16). On graduating *kandidat*, Chugaev was appointed Head of the Bacteriological Institute in Moscow. From 1904 to 1908, he was Professor at the Imperial Technical School (later the Moscow Higher Technical School), and from 1908 until his death, he held dual appointments as Professor at the St. Petersburg Technological Institute and St. Petersburg University. Chugaev's lasting contribution to organic chemistry is the Chugaev reaction, which is a pyrolytic *syn* elimination of xanthate esters to give alkenes, and his specific gravimetric reagent for nickel is still used (as the author can personally attest from his own experience!). In his later independent career, Chugaev focused his attention on the platinum metals and their complexes; in 1918, he became the founder and director of the Institute for the Study of Platinum and Other Noble Metals, a position he held until his untimely death at the age of 48.

Lev Aleksandrovich Chugaev
(Лев Александрович Чугаев, 1873-1922)

5.4 The Legacy of St. Petersburg

The organic chemistry community in St. Petersburg had been placed on a firm footing by Zinin, Voskresenskii, and Butlerov, by their student Menshutkin, and by Beilstein. By the end of the nineteenth century, St. Petersburg had become a major center for organic chemistry, but it is remarkable that the imperial capital still lagged behind Kazan' in terms of the number of its graduates occupying Chairs of Chemistry at universities in the empire.

5.4.1 The legacy of Butlerov and Mendeleev

During his years at St. Petersburg, Butlerov gathered a large group of students around him, although only a few of these achieved eminence on their own. One such was Aleksei Yevgrafovich Favorskii, the uncle of Prilezhaev, whom we discussed earlier, and who himself became mentor to important chemists. Although he did not make any major contributions to organic chemistry himself, Mendeleev did mentor students who did.

5.4.1.1 Aleksei Yevgrafovich Favorskii

Aleksei Yevgrafovich Favorskii (Алексей Евграфович Фаворский, 1860–1945) was born in Pavlovo, in the Nizhni Novgorod province, and in 1878 he entered St. Petersburg University, where he joined Butlerov's laboratory under the direction of Butlerov's long-time Assistant, Mikail Dmitrievich Lvov.[13] He graduated from the university with the degree of *kandidat* in 1883. In 1890, he successfully defended his M. Chem. dissertation, "On the question of the mechanism of isomerization in the field of unsaturated hydrocarbons" [80], in which he disclosed the base-catalyzed isomerization of ethylacetylene to dimethylacetylene by alkoxide bases at high temperature. Five years later, he successfully defended his Dr. Chem. dissertation, "Investigations of isomeric transformations in the fields of carbinol compounds, chlorinated ketones, and halogenated oxides."

In 1896, he was appointed Professor at the St. Petersburg Technological Institute, and in 1908 he moved to St. Petersburg University as Professor of Chemistry, a position he held until 1929. He was elected a Corresponding Member of the USSR Academy of Sciences in 1921, and in 1929 he was elected a full Academician. The same year, he was appointed Director of the Institute of Organic Chemistry of the USSR Academy of Sciences. He held this position until 1945.

Aleksei Yevgrafovich Favorskii
(Алексей Евграфович Фаворский,
1860-1945)

---

[13] Mikhail Dmitrievich L'vov (Михаил Дмитриевич Львов, 1846–1899) studied under Butlerov at Kazan' and St. Petersburg. At St. Petrsburg, he lectured at both St. Petersburg University and at the St. Petersburg Technological Institute. A long-term Assistant to Butlerov, he edited Butlerov's *Introduction to the Study of Organic Chemistry*, the first textbook based entirely on structural theory.

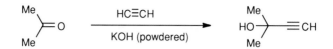

Fig. 5.17 The Favorskii-Babayan synthesis of acetylenic alcohols

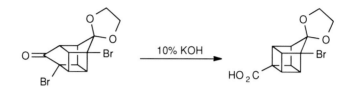

Fig. 5.18 The Favorskii rearrangement as used in Eaton's total synthesis of cubane

Favorskii was honored during his lifetime with an honorary membership of the Société Chimique de France, and was awarded the State Prize of the USSR (1941), four Orders of Lenin, the Order of the Red Banner of Labor, as well as being named a Hero of Socialist Labor (1945). The A. E. Favorskii Institute of Chemistry located in the Siberian Division of the Russian Academy of Sciences is named for him.

Favorskii's name is associated with two reactions: the Favorskii-Babayan synthesis of acetylenic alcohols (Fig. 5.17) [81], and the Favorskii rearrangement of α-haloketones (Fig. 5.18), promoted by base [82]. The Favorskii rearrangement was a key reaction in Eaton's total synthesis of the Platonic hydrocarbon, cubane [83], and the Favorskii-Babayan reaction provided the basis for the production of synthetic rubber by Russia during World War 2.

5.4.1.2 Vyacheslav Yevgen'evich Tishchenko

Although he began his research career with Butlerov, Vyacheslav Yevgen'evich Tishchenko (Вячеслав Евгеньевич Тищенко, 1861–1941) was predominantly associated with Mendeleev. He was born and educated in St. Petersburg, where he entered the university, and graduated *kandidat* in 1884, after spending 2 years in Butlerov's laboratory. During his student years, he met and married the sister of Aleksei Favorskii. On graduating, he took a position as a lecturer and laboratory assistant to Mendeleev, and worked in his laboratory for several years. He defended his M. Chem. dissertation, "On the action of amalgamated aluminum on alcohols" [84] in 1899, and his Dr. Chem. dissertation, "On the action of aluminum alcoholates on aldehydes. Ester condensation as a new type of condensation of aldehydes" [85] in 1906. His Dr. Chem. dissertation contained the

Fig. 5.19 The Tishchenko
reaction

$$Ph-CHO \xrightarrow{\text{Al(O-}t\text{-Bu)}_3} Ph-\overset{O}{\underset{O-\diagup}{\diagdown}}Ph$$

disproportionation reaction that now bears his name (Fig. 5.19) [86]. The analogous reaction of aldehydes with sodium alkoxides had been reported by Ludwig
Claisen in 1887 [87].

Vyacheslav Yevgen'evich Tis h-chenko
(Вячеслав Евгеньевич Тищенко, 1861-1941)

In 1891, Tishchenko was appointed Lecturer at St. Petersburg, and began to
lecture on analytical chemistry and chemical technology. His interests in chemical
technology made him the logical choice to send to the Exposition in Paris (1900)
and the Chicago World's Fair (1893) to report on the chemical technology
exhibited there. He became Professor of Chemistry at the Women's Medical
Institute in 1901, while still teaching at the university, and, in 1934, he became
Director of a Research Institute at Leningrad State University. Tishchenko was
elected a Corresponding Member of the USSR Academy of Sciences in 1928, and
a full Academician in 1935. In addition to his chemical work, Tishchenko spent a
considerable amount of time and energy compiling biographical information about
his favorite teacher, Mendeleev.

5.4.2 Favorskii's Legacy

5.4.2.1 Vladimir Nikolaevich Ipatieff

Vladimir Nikolaevich Ipatieff (Владимир Николаевич Ипатьев, 1867–1952) was born in Moscow, and educated at the Mikhail Military Academy, from which he graduated in 1892. After graduating, Ipatieff became an instructor at the Mikhail Artillery Academy, with the rank of captain. Needing a dissertation within 3 years of his appointment, he approached Favorskii, who had just graduated M. Chem. and been appointed Assistant Professor at St. Petersburg, to supervise his efforts. In 1995, Ipatieff presented his dissertation, which led to two publications [88], and his promotion to Assistant Professor. In 1896, he was sent to Munich to study with Baeyer. His first publication there came quickly, and carried both his name and Baeyer's [89]. In Baeyer's laboratory, he met Willstätter and Gomberg, with whom he remained life-long friends. Before his return to Russia in 1897, he traveled through Germany and France, visiting military institutions and discussed ballistics. In 1898, he attended the Second International Congress on Pure and Applied Chemistry, in Vienna, and he also wrote his dissertation for promotion, which he successfully defended at St. Petersburg [90]. This resulted in him being promoted to Professor of Chemistry and Explosives, the first full professor at the Military Academy.

Vladimir Nikolaevich Ipatieff
(Владимир Николаевич Ипатьев,
1867-1952), in uniform as Lieutenant General

The new century also saw a change in Ipatieff's research emphasis, as he began what became his most famous work: the investigation of catalytic reactions under high temperature and pressure. In 1908, he was invited to submit his dissertation for the degree of Dr. Chem. to St. Petersburg University; he successfully defended it and passed the associated examinations. It was necessary for him to obtain permission to submit his dissertation because, as an individual who had not graduated from the Gymnasium, he was not eligible for a higher degree. However, the university statute of 1884 permitted chemists who had attained an international stature to submit a dissertation for the degree, provided that the permission of the Ministry of Education was obtained. Four years after becoming Dr. Chem., he became Emeritus Professor at the Mikhail Artillery Academy. Ipatieff's military career continued apace during this time: in 1904, he was promoted to Colonel, and in 1910 he became Major General. Four years later, he was promoted to Lieutenant General. In 1914, Ipatieff was elected a Corresponding Member of the Academy of Sciences, and in 1916 he became a full Academician.

The years 1914–1918 were years of turmoil for Russia, encompassing as they did both World War 1, and the October revolution. Ipatieff was retained by the Soviet government, and was chosen to convert Russia's wartime chemical industry to a peacetime basis. From 1921 to 1926 he served a Chairman of the Supreme Council of the National Economy; despite never becoming a member of the Communist Party, Ipatieff was considered a government official. In 1927, he founded the Institute of High Pressures in the Artillery Academy.

During the late 1920s, Ipatieff worked part-time in Germany, at the Bayerische Stickstoff Werke, with the understanding that Russia would have a share of ownership in any discoveries he made. In 1930, during the period of arrests of scientists, Ipatieff left Russia with his wife as delegate to a conference in Berlin. He never returned to his homeland. In Berlin, he met Dr. Gustav Egloff, of Universal Oil Products, in Chicago, who arranged for him to obtain the necessary visas to visit the United States. In Chicago, he was offered, and accepted, the position of Director of Chemical Research at UOP. Shortly thereafter, he began his long association with Northwestern University. In 1939, he was elected to the National Academy of Sciences of the United States.

Ipatieff's name is memorialized in the form of the Ipatieff Prize, an award that he endowed, and that is given every 3 years by the American Chemical Society to a chemist under the age of 40 who has made outstanding chemical experimental work performed in the field of catalysis and high pressures.

5.4.2.2 Ivan Nikolaevich Nazarov

Ivan Nikolaevich Nazarov
(Иван Николаевич Назаров, 1906-1957)

Ivan Nikolaevich Nazarov (Иван Николаевич Назаров, 1905–1957) was the first Russian organic chemist to obtain his education under the Soviet system, and is one of the few Russian scientists from that era to have his name widely known in the west. Nazarov was born in the village of Koshelevo, located between Moscow and Nizhni Novgorod. He studied under Favorskii at the K. A. Timiryazev Moscow Agricultural Academy, graduating with the degree of *kandidat* in 1931. In 1934, he received his Dr. Chem., and was appointed to the Institute of Organic Chemistry of the USSR Academy of Science as an Associate. He remained here until 1947, when he was appointed Professor at the Lomonosov Institute of Fine Chemicals in Moscow. In 1946, he was elected a Corresponding Member of the USSR Academy of Science, and in 1953 he was elected a full Academician.

As a student of Favorskii, it is not surprising that most of Nazarov's works are devoted to the chemistry of acetylene and acetylene derivatives—vinyl acetylene, in particular. It was while studying the isomerization of allyl vinyl ketone, itself obtained by acid-catalyzed dehydration/hydration of vinyl allyl carbinol, that Nazarov discovered the electrocyclization reaction that bears his name (Fig. 5.20) [91]. In this reaction, a divinyl ketone cyclizes under strongly acidic conditions to give a cyclopentenone derivative through an intermediate 2-hydroxyallyl cation.

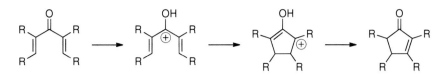

Fig. 5.20 The Nazarov cyclization

5.4.3 Odessa University and St. Petersburg: Nikolai Nikolaevich Sokolov

Nikolai Nikolaevich Sokolov
(Николай Никмолаевич Соколов, 1826-1877)

As outlined in Chap. 1, Odessa University was founded in 1865 as the Novorossiisk Imperial University, having been converted from the Richelieu Lycée. For 6 months, beginning in 1855, a decade prior to the formation of the university, Mendeleev taught mathematics and natural science at the Lycée while preparing his dissertation for the degree of M. Chem. He had been advised to move to southern Russia as treatment for tuberculosis. On his arrival in Simferopol in August, 1855, he found a city in the throes of the Crimean war, but he had the fortune to meet the eminent surgeon, Nikolai Ivanovich Pirogov (Николай

Иванович Пирогов, 1810–1881),[14] who confirmed that he did not, in fact, have the disease. Later, the same Pirogov played a prominent role in organizing the university.

The first occupant of the Chair of Chemistry at Odessa was Nikolai Nikolaevich Sokolov (1826–1877), whom we have already met as one of the mentors of Nikolai Menshutkin, at St. Petersburg. Sokolov was a gifted pedagogue who had enrolled at St. Petersburg University; since the cameral section in the Juridicial faculty was closed when he sought to enroll at the university, he enrolled in the natural science section, and studied chemistry under Voskresenskii, instead. He took his degree of *kandidat* in 1851, and was immediately awarded a *komandirovka*, which he used to study with Liebig in 1851, and with Gerhardt and Laurent in 1852. In 1859, he presented his dissertation for the degree of Dr. Chem. [92], and he was appointed Docent at St. Petersburg. In 1864, he moved to the Chair of Chemistry at Novorossiya University in Odessa but, in 1872 he returned to the capital to take the Chair of Chemistry at the St. Petersburg Institute of Forestry and Land Cultivation.

Along with Aleksandr Nikolaevich Engel'gardt (Александр Николаевич Енгельгардт, 1832–1893), a chemist who served in the St. Petersburg Arsenal, where he oversaw the casting of cannons, Sokolov founded the first (albeit short-lived) chemical journal in Russia (*Khimicheskii zhurnal*, The Chemical Journal), which published 24 issues in four volumes from 1859 to 1860. Both men were among the founders of the Russian Physical–Chemical Society.

In addition to his chemical teaching, Sokolov also served (anonymously) to provide accurate translations of the works of Charles Darwin, and to popularize Darwin's views in Russia. To quote Slavic scholar, Joakim Philipson [93]:

> The first accurate and complete Russian translation of one of Darwin's works appeared in 1860, when an article entitled "Geology" was published in the official organ of the Marine Authority, *Morskoi sbornik*,[38] which had recently been transformed into a broad literary-scientific journal with popular appeal. The anonymous translator of this particular article was most likely Nikolai Nikolaevich Sokolov (*Raikov* 1960:26). Although his field of research was chemistry and mineralogy, rather than biology, Sokolov played a role in the popularization of Darwinism in Russia through another anonymous article, entitled *Darvin i ego teoriia obrazovaniia vidov* - "Darwin and his theory on the formation of species", which was published in the broadly oriented journal "Library for reading" - *Biblioteka dlia chteniia* in 1861.

While at Giessen, Sokolov reported the detection of creatinine in the urine of calves and horses [94], and, with Adolph Strecker, he studied the decomposition of hippuric acid by nitric acid [95]. His later, independent work concerned the chemistry of hydroxy- and polyhydroxyacids, and led to the synthesis of glyceric acid by nitric acid oxidation of glycerine [96], and the synthesis of lactic acid from

---

[14] Pirogov was a prominent Russian scientist, doctor, and pedagogue, who was born in Moscow, and entered the university there at age 14. By 1836—at age 25—he had become Professor at the Imperial University of Dorpat. In 1840, he became Professor of Surgery at the Academy of Military Medicine in St. Petersburg. Pirogov pioneered the field surgery while serving with the army in the Crimea, and was one of the first surgeons in Europe to use ether as an anesthetic.

β-iodopropionic acid [97]. Sokolov died at 50 years of age, so his chemical output was quite small; it is his work as a teacher and pedagogue of science that remains his lasting legacy.

References

1. (a) Grignard V (1900). Sur quelques nouvelles combinaisons organometalliques du magnésium et leur application à des synthèses d'alcools et d'hydrocarbures. C R Hebd Séances Acad Sci, Ser C 130:1322-1324; (1901). Action des éthers d'acides gras monobasiques sur les combinaisons organomanésiennes mixtes. 132:336-338; Sur les combinaisons organomanésiennes mixtes. ibid 132:558-561; (1902). Action des combinaisons organomanésiennes sur les éthers α-cétoniques. ibid 134:849-851; (1901) Über gemischte Organomagnesiumverbindungen und ihre Anwendung zu Synthesen von Säuren, Alkoholen und Kohlenwasserstoffen. Chem Zentralbl pt. II 622-625; (1901). Sur les combinaisons organomanésiennes mixtes et leur application a des synthèses d'acides, d'alcohols et d'hydrocarbures. Ann Chim [vii] 24:433-490. (b) Tissier L, Grignard V (1901). Sur les composés organométalliques du magnésium. C R Hebd Séances Acad Sci, Ser C 132:835-837; Action des chlorures d'acides et des anhydrides d'acides sur les composés organo-metalliques du magnésium. ibid 683-685. (c) Grignard V, Tissier L (1902). Action des combinaisons organomanésiennes mixtes sur le trioxyméthylène: synthèses d'alcools primaires. C R Hebd Séances Acad Sci, Ser C 134:107-108; (1902). errata, ibid 134:1260.
2. For an early account of the resurgence of the Zaitsev-Vagner synthesis of alcohols in asymmetric form, see: Noyori R (1990). Chiral Metal Complexes as Discriminating Molecular Catalysts. Science, 248:1194-1199.
3. Arbuzov AE (1901). Ob allilmetilfenilkarbinoli [On Allylmethylphenylcarbinol]. Zh Russ Fiz-Khim O-va 33:38-45.
4. Arbuzov AE (1902). Primenenie v Kazanskoi laboratorii magniya dlya polucheniya tretichnykh spirtov po sposobu Zaitseva [The use of magnesium in the laboratory of Kazan for tertiary alcohols by the method of Zaitsev]. Zh Russ Fiz-Khim O-va 34:1.
5. For biographical information about Arbuzov in Russian, see: (a) Bogoyavlenskii AF, Aksenov, NN (1946). Aleksandr Erminingel'dovich Arbuzov. Kazan; (b) Grechkin NP, Kuznetsov VI (1977) Aleksandr Erminingel'dovich Arbuzov 1877-1968. Izd-vo Akad Nauk SSSR, Moscow. For material in English, see: (c) Unnamed biographer (1962). Aleksandr Erminingel'dovich Arbuzov. Russ Chem Bull 11:1625-1626.
6. Arbuzov AE (1905). O stroenii fosforistoi kisloty i ee proizbodnykh. [Concerning the structure of phosphorous acid and its derivatives]. M Chem Diss, Kazan'.
7. Arbuzov AE (1906). O stroenii fosforistoi kisloty i eya proizvodnykh. Glava I. Kratkii istoricheskii ocherk voprosa o stroenii fosforistoi kisloty i eya proizvodnykh [On the structure of phosphorous acid. Part I. A brief history of the question of the structure of phosphorous acid and its production]. Zh Russ Fiz-Khim O-va 37:161-187; Glava II. O poluchenii efirov fosforistoi kisloty obshchogo vida $P(OR)_3$ [Part II. On the formation of esters of phosphorous acid of general form $P(OR)_3$]. ibid 187-228; Glava III. O soediniyakh proizvodnykh trekhatnogo fosfora s odnogaloidnymi solyani medi [Part III. On the formation of compounds of trivalent phosphorus with copper monohalide salts]. ibid 37:293-319; Glava IV. Izomeratsiya i perekhod soedinenii trekhatomnogo fosfora v soedineniya pyatiatomnogo [Part IV. Isomerization and conversion of a compound of trivalent phosphorus into a pentavalent compound]. ibid 37:687-721; (1910). O protsessakh izomerizatsii v oblasti nekotorykh soedinenii fosfora. Stat'ya pervaya. [On the processes of isomerization in the field of certain compounds of phosphorus. First article.] ibid 42:395-420; O protsessakh izomerizatsii v oblasti nekotorykh soedinenii fosfora. Stat'ya vtoraya. [On the processes of

isomerization in the field of certain compounds of phosphorus. Second article]. ibid 42:549-561. (b) Arbusow A (1906). Chem Zentr II:1639; (1910) 453.

8. Arbuzov AE (1948). Kratkii ocherk razvitiya organicheskoi khimii v Rossii. [A short account of the development of organic chemistry in Russia]. Moscow-Leningrad.

9. Arbuzov AE (1914). O yavleniyakh kataliza v oblasti prevrashchenii nekotorykh soedinenii fosfora [On the phenomena of catalysis in the reactions of some phosphorus compounds]. Dr Chem Diss, Kazan'.

10. For a biography of Popov, see: Bykov GV (1956). Ocherk zhizni i deyatelnosti Aleksandra Nikiforovicha Popova [A Sketch of the Life and Work of Aleksandr Nikiforovich Popov]. In Trudy Instituta istorii estestvoznaniya i tekhniki [Proc Inst Hist Sci Technol] 12:200–245.

11. Popov AN (1865). Po povodu srodstva uglerodnogo atoma [Regarding the affinity of the carbon atom]. Kand Diss, Kazan'.

12. Popov AN (1869). Ob okislenii ketonov odnoatomnykh [On the oxidation of monoatomic ketones]. Dr Chem Diss, Kazan'.

13. Popoff A (1865). Ueber die Isomerie der Ketone. Z Chem n.s. 1:577–580.

14. Popoff A (1866). Sur l'isomérie des acétones. Bull Soc Chim Paris Nouv Sér 5:35-42; (1868). Ueber die Isomerie der Ketone. Justus Liebigs Ann Chem 145:283-292.

15. (a) Popoff A, Zincke T(1872). Bestimmung der Constitution von Alkoholradikalen durch Oxydation aromatischer Kohlenwasserstoffe. Ber Dtsch Chem Ges 5:384–387.

16. (a) Lewis DE (2010). Feuding Rule-Makers: Vladimir Vasil'evich Markovnikov (1838-1904) and Aleksandr Mikhailovich Zaitsev (1841-1910). A Commentary on the Origins of Zaitsev's Rule. Bull Hist Chem 35:115-124. (b) Lewis DE (2011). A. M. Zaitsev: Lasting contributions of a synthetic virtuoso a century after his death. Angew Chem Int Ed 50:6452-6458; A. M. Saytzeff: bleibendes Vermächtnis eines Virtuosen der Synthesechemie. Angew Chem 123:6580-6586.

17. For biographies of Wagner, see: (a) Bewad J, Brykner W, Goldsobel AG, Ertschikowsky A, Lagorio A, Lawrow W, Slowinski K, Wagner G jun. (1903). Georg Wagner. Ber Dtsch Chem Ges 35:4591-4613. (b) Sementsov A (1966). Egor Egorovich Vagner and his role in terpene chemistry. Chymia 11:151-155. (c) Starosel'skii NI, Inkulina EP (1977). Egor Egorovich Vagner 1849-1903. Izd-vo "Nauk," Moscow.

18. Vagner E, Zaitsev A (1874). Sintez dietilkarbinola, novogo izomera amil'nogo alkogolya [The synthesis of diethyl carbinol, a new isomer of amyl alcohol]. Zh Russ Khim O-va Fiz O-va 6:290-308.

19. (a) Vagner E (1876). Deistvie tsinketila na uksusnyi al'degid [The action of zinc ethyl on acetaldehyde]. Zh Russ Khim O-va Fiz O-va 8:37-40. (b) Vagner E (1884). Ob otnoshenii al'degidov k tsinkorganicheskim soedineniyam (obshchii sposob polucheniya vtorichnykh spirtov) [On the eaction of aldehydes with organozinc compounds (a general method for secondary alcohols)]. Zh Russ Fiz-Khim O-va 16:283-353.

20. Vagner E (1885). Sintez vtorichnikh spirtov i ikh okislenie [The synthesis of secondary alcohols and their oxidation]. M Chem Diss, St Petersburg.

21. Wagner G (1885). Zur Oxydation der Ketone. Ber Dtsch Chem Ges 18:2266-2269.

22. Saytzeff A (1885). Ueber die Oxydation der Oelsäure mit Kaliumpermanganat in alkalischer Lösung. J Prakt Chem 31:541-542; (1886). Untersuchungen aus dem chemischen Laboratorium von Prof. Alexander Saytzeff zu Kasan. 26. Ueber die Oxydation der Oel- und Elaïdinsäure mit Kaliumpermanganat in alkalischer Lösung. ibid 33:300-318.

23. Tanatar S (1879). Ueber Bioxyfumarsäure. Ber Dtsch Chem Ges 12:2293-2298; (1880). Reindarstellung der Bioxyfumarsäure. ibid 13:159; Trioxymaleïnsäure. ibid 13:1383-1388.

24. (a) Kekulé A, Anschütz R (1880). Ueber Tanatars Bioxyfumarsäure Ber Dtsch Chem Ges 13:2150-2152; (1881). Ueber Tanatar's Trioxymaleinsäure. ibid 14:713-717; (b) Fittig R (1888). Ueber das Verhalten der ungesättigten Säuren bei vorsichtiger Oxydation. Ber Dtsch Chem Ges 21:919-921. (c) Baeyer A (1888). Ueber die Constitution des Benzols. Justus Liebigs Ann Chem 245:103-190.

25. Wagner G (1888). Ueber die Oxydation der Olefine und der Alkohole der Allyalalkoholreihe. Ber. Dtsch. Chem. Ges. 21:1230-1240; Ueber die Oxydation der Kohlenwasserstoffe, C_nH_{2n-2}.

ibid 21:3343-3346; Zur Oxydation ungesättigter Verbindungen. ibid 21:3347-3355; Zur Frage über die Betheiligung des Wassers an der Oxydation ungesättigter Verbindungen. ibid 21:3356-3360.

26. Vagner E (1888). K reaktsii okisleniya nepredel'nykh uglerodistykh soedinenii [Oxidation reactions of unsaturated carbon compounds]. Dr Chem Diss, Warsaw.

27. (a) Wagner G (1890). Ueber Camphenglycol und den vieratomigen Alkohol aus Limonen. Ber Dtsch Chem Ges 23:2307-2318; (1894). Zur Oxydation cyclischer Verbindungen. ibid 27:1636-1654.

28. Wagner G (1890). Ueber Camphenglycol und den vieratomigen Alkohol aus Limonen. Ber Dtsch Chem Ges 23:2307-2318.

29. (a) Wagner G, Ertschikowsky G (1896). Zur Oxydation des Pinens. Ber Dtsch Chem Ges 29:881. (b) Wagner G, Ginzberg A (1896). Zur Constitution des Pinens. Ber Dtsch Chem Ges 29:886. (c) Ginzberg A, Vagner E (1896). Stroenie pinena [The structure of pinene]. Zh Russ Fiz-Khim O-va 28:494-501. (d) Wagner G, Slawinski K (1899). Zur Constitution des Pinens. Ber Dtsch Chem Ges 32:2064.

30. Vagner EE (1899). O stroenii kamfena (soobshchenie) [About the structure of camphene (communication)]. Zh Russ Fiz-Khim O-va 31:680-684.

31. Wagner G, Brykner W (1900). Bornylen, ein neues Terpen. Ber Dtsch Chem Ges 33:2121-2125.

32. Wagner G, Brickner W (1899). Ueber die Beziehung der Pinenhaloïdhydrate zu den Haloïdanhydriden des Borneols. Ber Dtsch Chem Ges 32:2302-2325.

33. (a) Meerwein H, van Emster K (1920). Untersuchungen in der Camphen-Reihe, I.: Über den Reaktionsmechanismus der Isoborneol $\rightleftharpoons$ Camphen-Umlagerung. Ber Dtsch Chem Ges 53:1815-1829. (b) Meerwein H, van Emster K, Jousson J (1922). Über die Gleichgewichts-Isomerie zwischen Bornylchlorid, Isobornylchlorid und Camphen-chlorhydrat. Ber Dtsch Chem Ges.55:2500-2528. (c) Meerwein H, Wortmann R (1924). II. Über das Campherdichlorid. Justus Liebigs Ann Chem 435:190-206.

34. Ipatieff VN (1946). The life of a Chemist. Memoirs of Vladimir N. Ipatieff. Stanford University Press, p. 101.

35. Baeyer A (1896). Ortsbestimmungen in der Terpenreihe. Ber Dtsch Chem Ges 29:3-26.

36. Meerwein H (1914). Über den Reaktionsmechanismus der Umwandlung von Borneol in Camphen; [Dritte Mitteilung über Pinakolinumlagerungen.] Justus Liebigs Ann Chem 405:129-175.

37. For a biographical sketch of Prilezhaev, see: Akhrem, AA, Prilezhaeva EN, Meshcheryakov AP (1951). Zhizn' i deyatel'nost' N. A. Prilezhaeva [The life and works of N. A. Pilezhaev]. Zh Obshch Khim 21:1925-1928.

38. Prileschajew N (1909). Oxydation ungesättigter Verbindungen mittels organischer Superoxyde. Ber Dtsch Chem Ges 42:4811-4815.

39. Prilezhaev NA (1912). Organicheskie perekisi i ikh primenenie dlya okisleniya nepredel'nykh soeninenii [Organic peroxides and their application to the oxidation of unsaturated compounds]. M Chem Diss, Warsaw.

40. For a biography of Reformatskii, see: Semenzow A (1935). Sergius Reformatsky. Ber Dtsch Chem Ges 68:61A.

41. Reformatskii SN (1882). Issledovanie uglevodoroda $C_{10}H_{18}$, poluchaemogo iz allildipropilkarbinola [A study of the hydrocarbon $C_{10}H_{18}$ produced from allyl dipropyl carbinol]. Kand Diss, Kazan'.

42. Reformatsky S (1887). Ueber die Darstellung einiger mehratomiger Alkohole und ihrer Derivate mittelst unterchloriger Säure. vorläufige Mittheilung. J Prakt Chem 31:318-319.

43. Reformatskii SN (1889). Predel'nye mnogoatomnye alkogoli [The limit of polyhydric alcohols]. M Chem Diss, Kazan'.

44. Reformatskii SN (1890). Deistvie smesi tsinka i monoxloruksusnogo efira naketony i al'degidy [The action of a mixture of zinc and monochloroacetic ester on ketones and aldehydes]. Dr Chem Diss, Warsaw.

45. For reviews of the Reformatskii reaction, see: (a) Shriner RL (1942). The Reformatsky reaction. Org React 1:1-37. (b) Diaper DGM, Kukis A. (1959). Synthesis of alkylated alkanedioic acids. Chem Rev 59:89-178. (c) Rathke MW (1975). The Reformatsky reaction. Org React 22:423-460. (d) Ocampo R, Dolbier, WR Jr. (2004). The Reformatsky reaction in organic synthesis. Recent Advances. Tetrahedron 60:9325-9374.

46. Reformatsky S (1887). Neue Synthese zweiatomiger einbasischer Säuren aus den Ketonen. Ber Dtsch Chem Ges 20:1210-1211.

47. (a) Reformatsky S (1895). Neue Darstellungsmethode der αα-Dimethylglutarsäure aus der entsprechenden Oxysäure. Ber Dtsch Chem Ges 28:3262-3265; (1895). Die Einwirkung eines Gemenges von Zink und Bromisobuttersäureester auf Isobutyraldehyd. Synthese der secundären β-Oxysäuren. ibid 28:2842-2847; (1897). Ueber den Zerfall der β-Monooxysäuren. J Prakt Chem 54:477-481.

48. Reformatsky S, Plesconosoff B (1895). Die Einwirkung eines Gemenges von Zink und Bromisobuttersäureester auf Aceton. Synthese der Tetramethyläthylenmilchsäure. Ber Dtsch Chem Ges 28:2838-2841.

49. Dem'yanov NYa (1886). O dekstrinakh [On dextrin]. Kand Diss, Moscow.

50. Gustavson G (1880). Ueber in Gegenwart von Aluminiumchlorid und Aluminiumbromid verlaufende Reactionen. Ber Dtsch Chem Ges 13:157-159; (1886). Die Einwirkung von Bromaluminium auf Aethylen und die Bromide der „Grenzalkohole." J Prakt Chem 34:161-177; (1888) Ueber die Produkte der Einwirkung von Chlor-aluminium auf Acetylchlorid. ibid 37:108-110; (1890). Ueber die Ursachen der Reactionen in Gegenwart von Chlor- und Bromaluminium. ibid 42:501-507.

51. Gustavson G (1887). Ueber eine neue Darstellungsmethode des Trimethylens. J Prakt Chem 36:300-303; (1890). Ueber die Einwirkung des Chlors auf Trimethylen. ibid 42:495-500; (1891). Ueber die Reaktionsfähigkeit des Monochlortrimethylens und einiger verwandten Verbindungen. ibid 43:396-402.

52. Gustavson G, Demjanoff N (1888). Ueber die Darstellung und Eigenschaften des allens. J Prakt Chem 38:201-207; (1889). Ueber die Bromide des Pentamethylens und Tetramethylens. ibid 39:542-543.

53. Dem'yanov NYa (1895). O deistvii azotistoi kisloty na tri-, tetra- i pentametilendiaminy i o metiltrimetilene [On the action of nitrous acid on tri-, tetra and pentamethylenediamines and on methyltrimethylene]. M Chem Diss, St. Petersburg.

54. Dem'yanov NYa (1899). O deistvii azotnogo angidrida i azotnovatoi okisi na uglevodorody etilenovogo ryada [On the action of nitric anhydride and nitrogen oxides on hydrocarbons of the ethylenic series]. Dr Chem Diss, Moscow.

55. (a) Dem'yanov NYa, Lushnikov M (1903). O deistvii azotistoi kisloty na tetrametilenilmetilaminy i o metiltrimetilene [On the action of nitrous acid on tetramethylenylmethylamine]. Zh Russ Fiz-Khim O-va 35:26-42. (b) Demjanow NJ (1907). Die Ringerweiterung bei den cyclischen Aminen mit der Seitenkette $CH_2.NH_2$. Über den Alkohol aus dem Amin

$$\begin{array}{c} CH_2 \\ | \\ CH_2 \end{array} \!\!\! >\!\! CH.\ CH_2.NH_2$$

Ber Dtsch Chem Ges 40:4393-4397. (c) Demjanow NJ, Dojarenko M (1908). Über einige Umwandlungen des Cyclobutanols. Ber Dtsch Chem Ges 41:43-46.

56. Demjanow NJ (1907). Die Umwandlung des Tetramethylenringes in den Trimethylenring. Ber Dtsch Chem Ges 40:4961-4963.

57. For historical perspectives on the polemics surrounding the non-classical norbornyl cation, see: (a) Davenport DA (1987). On opinion in good men. An oblique tribute to H. C. Brown. Aldrichimica Acta 20:25-27. (b) Weininger SJ (2000). "What's in a name?" From designation to denunciation—the non-classical cation controversy. Bull Hist Chem 25:123-131.

58. Tiffeneau M, Weill P, Tchoubar B (1937). Isomérisation de l'oxyde de méthylène cyclohexane et désamination de l'aminoalcool correspondant en cycloheptanone. Comptes Rend 205:54-56.

59. Kizhner NM (1895). Aminy i gidraziny polimetilenogo ryada, metody ix obrazovaniya i prevrashchenniya [Amines and hydrazines of the polymethylene series, methods for their generation and transformations]. M Chem Diss, St Petersburg.

60. Kizhner NM (1900). O deistvii okisi serebra i gidroksilamina na bromaminy. O stroenii geksagidrobenzola [On the action of silver oxide and hydroxylamine on bromamine. On the structure of hexahydrobenzene]. Dr Chem Diss, Moscow.

61. Kizhner NM (1912). Razlozhenie alkilidengidrazonov. XXXIV. Preobrazovanie furfural'degida v a-metilfurane. [Decomposition of alkylidenehydrazines. XXXIV. Transformation of furfuraldehyde into α-methylfuran]. Zh Russ Fiz-Khim O-va 43: 1563-1565; 582; (1912). Chem Abstr 6:347.

62. Wolff L (1912). Chemischen Institut der Universität Jena: Methode zum Ersatz des Sauerstoffatoms der Ketone und Aldehyde durch Wasserstof. Justus Liebigs Ann Chem 394:86-108.

63. Huang Minlon (1946). A simple modification of the Wolff-Kishner reduction. J Am Chem Soc 68:2487-2488.

64. Tschtschibabin AE (1907). Über einige phenylierte Derivate des p, p-Ditolyls. Ber Dtsch Chem Ges 40:1810-1819.

65. Chichibabin AE (1903). O produktakh deistvii galoidnykh soedinenii na piridin i khinolin [On the porducts of the action of halogen compounds on pyridine and quinoline]. M Chem Diss, Moscow.

66. Chichibabin AE (1912). Issledovaniya po voprosu o trekhatomnom uglerode i stroenii prosteishikh okrashennykh proizvondykh trifenilmetana [Studies on triatomic carbon and the structure of the simplest colored derivatives of triphenylmethane]. Dr Chem Diss, St. Petersburg.

67. Tschitschibabin AE (1904). Eine neue allgemeine Darstellungsmethode der Aldehyde. Ber Dtsch Chem Ges 37:186-188; Ueber den Hexahydro-m-toluylaldehyd. ibid 850-853.

68. Chichibabin AE (1906). K voprosu o sposobnosti etoksil'noi gruppy k zameshcheniyu na radikaly. Sintez atsetalei al'degidokislot i gomologicheskikh etoksiakrilovykh kislot [On the question of the ability of the ethox group to be substitute by radicals, synthesis of acetals of aldehyde-acid and homologous ethoxyacrylic acids]. Zh Russ Fiz-Khim O-va 37:327-343.

69. (a) Chichibabin AE, Zeide OA (1914). Zh Russ Fiz-Khim O-va 46:1216. (b) Tshitschibabin AE (1924). Über Kondensationen der Aldehyde mit Ammoniak zu Pyridinbasen. J Prakt Chem 107:122-128. (c) Tschitschibabin AE, Oparina MP (1924). Über die Synthese des Pyridins aus Aldehyden und Ammoniak. J Prakt Chem 107:154-158.

70. Zelinskii ND (1889). K voprosu ob izomerii v tiofenovom ryadu [On the question of isomerism in the thiophene series]. M Chem Diss, Odessa.

71. Zelinskii ND (1891). Issledovanie yavlenii stereoizomerii v ryadakh predel'nykh uglerodistykh sedinenii [Investigations of the phenomenon of stereoisomerism in saturated carbon compounds]. Dr Chem Diss, Odessa.

72. (a) Hell C (1882). Ueber eine neue Bromirungsmethode organischer Säuren. Ber Dtsch Chem Ges 14:891-893. (b) Volhard J (1887). 4) Ueber Darstellung α-bromirter Säuren. Justus Liebigs Ann Chem 242:141-163. (c) Zelinsky N (1887). Ueber eine bequeme Darstellungsweise von α-Brompropionsäureester. Ber Dtech Chem Ges 20:2026.

73. Zelinsky ND, Stadnikoff G (1908). Ein Beitrag zur Synthese des Alanins und der α-Aminobuttersäure. Ber Dtsch Chem Ges 41:2061-2063.

74. Zelinsky N (1911). Über Dehydrogenisation durch Katalyse. Ber Dtsch Chem Ges 44:3121-3125; (1912). Über die selektive Dehydrogenisations-Katalyse. ibid 45:3678-3682.

75. Zelinsky N (1924). Über die Kontakt-Kondensation des Acetylens. Ber Dtsch Chem Ges 57:264-276.

76. Chugaev LA (1903). Issledovaniya v oblasti terpenov i kamfory [Investigations in the field of terpenes and camphors]. M Chem Diss, Moscow.

77. Tschugaeff L (1899). Ueber eine neue Methode zur Darstellung ungesättigter Kohlenwasserstoffe. Ber Dtsch Chem Ges 32:3332-3335; (1900). Ueber die Umwandlung von Carvon in Limonen. ibid 33:735-736; Ueber das Thujen, ein neues bicyclisches Terpen ibid 33:3118-3126.

78. Chugaev LA (1906). Issledovaniya b oblasti kompleksnykh soedinenii [Investigations in the field of complex compounds]. Dr Chem Diss, Moscow.

79. Tschugaeff L (1905). Ueber ein neues, empfindliches Reagens auf Nickel. Ber Dtsch Chem Gres 38:2520-2522.

80. Favorskii AE (1891). O voprosu o mekhanizme izomerizatsii v ryadakh nepredel-nykh uglevodorodov [On the question of the mechanism of isomerization in the field of unsaturated hydrocarbons]. M Chem Diss, St. Petersburg.

81. Favorskii A (1905). Deistvie edkogo kali na cmes' ketonov s fenilatsetilenom [The effect of caustic potash on a mixture of ketones and phenylacetylene]. Zh Russ Fiz-Khim O-va 37:643-645. This paper is followed by a series of short papers by Favorskii's students, Skosarevskii M (645-647), Borka I (647-650; 650-652), Neverovich N (652-654), Bertonda E. (655-656, 657-659), Kotkovskii ya (659-661) all covering variants of this reaction.

82. (a) Faworsky A (1913). Über die Einwirkung von Phosphorhalogenverbindungen auf Ketone, Bromketone und Ketonalkohole. J Prakt Chem 88:641-698. (b) Favorskii AE, Bozhovskii VN (1914) Zh Russ Fiz-Khim O-va 46:1097 [Chem Abstr (1915) 9:1900].

83. Eaton PE, Cole TW (1964). Cubane. J Am Chem Soc 86:3157-3158.

84. Tishchenko VE (1899). O deistvii amal'gamirovannogo alyuminiya na alkogoli [On the action of amalgamated aluminum on alcohols]. M Chem Diss, St Petersburg.

85. Tishchenko VE (1906). O deistvii alkogolyatov alyuminiyana al'degidy. Slozhhoefirnaya kondensatsiya kak novyi vid uplotneniya al'degidov [On the action of aluminum alcoholates on aldehydes. Ester condensation as a new type of condensation of aldehydes]. Dr Chem Diss, St Petersburg.

86. (a) Tishchenko VE (1906). Deistvie alkogolyatov alyuminiya na aldegidy [The action of aluminum alcoholates on aldehydes.]. Zh Russ Fiz-Kihm O-va 37:382-418. O deistvii alkogolyatov alyuminiya na aldegidy. Slozhnoefirnaya kondensatsiya kak novyi vid uplotneniya aldegidov [On the action of aluminum alcoholates on aldehydes. Ester condensation as a new type of condensation of aldehydes]. ibid 38: 482-540.

87. Claisen L (1887). Ueber die Einwirkung von Natriumalkylaten auf Benzaldehyd. Ber Dtsch Chem Ges 20:646-650.

88. (a) Ipatiew W (1895). Ueber die Einwirkung von Bromwasserstoff auf Kohlenwasserstoffe der Reihe C_nH_{2n-2}. J Prakt Chem 53:145-168; Ueber die Einwirkung von Brom auf tertiäre Alkohole der Reihe $C_nH_{2n+2}O$. ibid 53:257-287.

89. Baeyer A (1896). Orsbeststimmungen in der Terpenreihe. Baeyer A, Ipatiew W (1896). Ueber die Caronsäure. Ber Dtsch Chem Ges 29:2796-2802.

90. Ipat'ev VN (1898). [Allene Hydrocarbons, the Reaction of Nitrosyl Chloride on Organic Compounds with a Double Bond and Nitrosates]. M Chem Diss, St. Petersburg.

91. Nazarov IN, Zaretskaya II (1941). [Acetylene derivatives, XVII. Hydration of hydrocarbons of the divinylacetylene series]. Izv Akad Nauk SSSR Ser Khim 211–224.

92. Sokolov NN (1859). O vodorode v organicheskikh soedineniyakh [On hydrogen in organic compounds]. Dr Chem Diss, St Petersburg.

93. Philipson J (2008). The Purpose of Evolution: the 'struggle for existence' in the Russian-Jewish press 1860-1900. Acta Universitatis Stockholmiensis, Doct Diss p. 73.

94. (a) Socoloff N (1851). Notiz über das Vorkommen des Kreatinins im Kälberharn. Justus Liebigs Ann Chem 80:114-117. (b) Socoloff N (1852). Sur l'existence de la Créatinine dans l'Urine de veau. Ann Chim 34:492-493. (c) Socoloff N (1852). Sur l'existence de la créatinine dans l'urine de veau. J Pharm Chim 21:443-444. (d) Socoloff N (1851). Notiz über die Anwesenheit des Kreatinins in dem Pferdeharn. Justus Liebigs Ann Chem 78:243-246. (e) Sokolov NN (1858). O prisutstvii kreatinina v moche loshadi i telenka [On the presence of creatine in the urine of horses and a calf]. Bull St Pétersb Acad Sci 61 col 369-378.

95. (a) Socoloff N, Strecker A (1851) Untersuchung einiger aus der Hippursäure entstehenden Producte. Justus Liebigs Ann Chem 80:17-43. (b) Socoloff N, Strecker A (1852). Recherches sur quelques produits dérivés de l'Acide hippurique. Ann Chim 34:232-240. (c) Socoloff N, Strecker A (1852). Recherches sur quelques produits dérivés de l'acide hippurique. J Pharm Chim 1852 21 237-240.

96. (a) Socoloff (1858). Sur l'oxydation de al Glycérine par l'acide nitrique. Ann Chim 54:95-96. (b) Sokolof N (1858). Ueber die Oxydation des Glycerins durch Salpetersäure. Erdmanns J Prakt Chem 1858 75 302-313. (c) Socoloff N (1858) Ueber die Oxydation des Glycerins durch Salpetersäure. Justus Liebigs Ann Chem 101:95-108.

97. Socoloff N (1869). Ueber die Milchsäure aus β-Jodpropionsäure. Justus Liebigs Ann Chem 150:167-187.

Photograph Sources and Credits

Most of the photographs in this book are of sufficient age that they are now in the public domain. Other photographs used in this book have been placed in the public domain by their creators. The photographs were obtained from the following sources:

The Kunstkamera in St. Petersburg, first location of the Imperial Academy of Sciences.

©2006 Sergey Barichev. http://en.wikipedia.org/wiki/File:Kunstkamera_(Saint-Petersburg).jpg Downloaded under the terms of the GNU Free Documentation License, version 1.2.

Moscow University, 1786. Downloaded from the official website of Lomonosov Moscow State University, http://www.msu.ru/en/info/history.html

The St. Petersburg Medical-Surgical Academy (now the Army Medical Academy). Downloaded from http://ru.wikipedia.org/wiki/Файл:Vma_by_Bulla.jpg

Dorpat University in 1860 (Lithograph *Die Universitätsgebäude* by Louis Höflinger, 1860). Downloaded from http://en.wikipedia.org/wiki/University_of_Tartu

The Grand Courtyard of Vilna University and the Church of St, John. ("Album Wileński" J. K. Wilczińskiego, 1850). Downloaded from http://en.wikipedia.org/wiki/Vilnius_University

Khar'kov Imperial University, prior to 1918 downloaded from the official website of V. N. Karazin Kharkiv National University, http://www.univer.kharkov.ua/en/general/our_university/history

Kazan' Imperial University, *ca* 1815. Dowloaded from the official website of the Museum of History of Kazan University, http://www.ksu.ru/miku/eng/ekskurs/istoki/s2.php

The Twelve Collegia Building as it appears in a 1753 lithograph. Downloaded from http://en.wikipedia.org/wiki/File:Twelvecollegia.jpg

Main gate to the University of Warsaw. A view from the Institute of Philosophy, May 2007. Downloaded from http://en.wikipedia.org/wiki/File:Bramauw.jpg This image has been released to the public domain by the author.

D. E. Lewis, *Early Russian Organic Chemists and Their Legacy*,
SpringerBriefs in History of Chemistry, DOI: 10.1007/978-3-642-28219-5,
© The Author(s) 2012

Biographical Sketch of the Author

David E. Lewis was born and educated in South Australia, where he took a first-class degree of B.Sc. (Hons.), and a Ph.D. degree in organic chemistry from the University of Adelaide under R. A. Massy-Westropp, working in the area of natural products structure determination. After completing his Ph.D. research, Lewis moved to the United States in December, 1976. At Arkansas, he worked as Research Associate at the University of Arkansas under Leslie B. Sims and Arthur Fry, working in the area of kinetic isotope effects and physical organic chemistry. Two years later, he was appointed as Lecturer in Chemistry at Arkansas. In 1980, he moved to the University of Illinois at Urbana-Champaign as Visiting Assistant Professor, working with Kenneth L. Rinehart, Jr., in the area of organic synthesis. In 1981, he moved to his first tenure-track position as Assistant Professor of Chemistry at Baylor University in Waco, Texas; he was promoted to Associate Professor in 1988. In 1989, he moved to South Dakota State University, in Brookings, South Dakota, as Associate Professor of Chemistry; he was promoted to Professor of Chemistry in 1993. In 1997, he moved to the University of Wisconsin-Eau Claire as Professor and Chair of Chemistry; he stepped down as Chair in 1999.

Lewis' research interests in organic synthesis may be broadly defined as applied organic chemistry, ranging from applications of fluorescent dyes in engineering and biology, to the synthesis of new compounds with potentially useful biological activity. He has had a two-decade interest in the history of organic chemistry in pre-revolutionary Russia, and it is this interest that has led to this volume. Lewis is a long-time member of the American Chemical Society, where he has served as Chair of the Division of the History of Chemistry, and is a Fellow of the Royal Australian Chemical Institute. He is the holder of 18 U.S. and international patents, and has published over 65 research articles, book chapters, and books.

Index

D. E. Lewis, *Early Russian Organic Chemists and Their Legacy*,
SpringerBriefs in History of Chemistry, DOI: 10.1007/978-3-642-28219-5,
© The Author(s) 2012